FINANCING PARKING FACILITIES

FINANCING PARKING FACILITIES

E. Carlton Heeseler
Financial Consultant

VNR VAN NOSTRAND REINHOLD
New York

Softcover reprint of the hardcover 1st edition 1991

Library of Congress Catalog Card Number 90-22782
ISBN-13: 978-1-4684-6529-7

Published by Van Nostrand Reinhold
115 Fifth Avenue
New York, New York 10003

Chapman and Hall
2-6 Boundary Row
London, SE1 8HN

Thomas Nelson Australia
102 Dodds Street
South Melbourne 3205
Victoria, Australia

Nelson Canada
1120 Birchmount Road
Scarborough, Ontario M1K 5G4, Canada

16 15 14 13 12 11 10 9 8 7 6 5 4 3 2 1

Library of Congress Cataloging-in-Publication Data
Heeseler, E. Carlton.
Financing parking facilities / E. Carlton Heeseler.
p. cm.
Includes bibliographical references and index.
ISBN-13: 978-1-4684-6529-7 e-ISBN-13: 978-1-4684-6527-3
DOI: 10.1007/978-1-4684-6527-3
1. Automobile parking—United States—Finance. I. Title.
HE355.3.P37H44 1991
388.4'74—dc20 90-22782
CIP

CONTENTS

PREFACE

The level of knowledge concerning the financing of parking facilities is high among private parking operators and owners. Internal growth and new projects would not occur, if parking companies were not thoroughly familiar with the ability to finance new projects and acquisitions, at advantageous terms. Understandably, each private parking company jealously guards not only its source of borrowed or equity funds but also the methodology of the financing used.

The vast majority of new parking projects, however, are planned, financed, and owned by developers, industrial companies, and municipalities. While each of these categories of parking facility owners have knowledge of their respective methods and sources of financing for general corporate or municipal purposes, the level of expertise concerning specifically the financing of parking facilities is sadly lacking. If for no other reason, both private corporations and municipalities frequently include the cost of a new parking project as part of an overall financing package. Developers usually include the cost of required parking facilities within the overall project development financing. Municipalities frequently include the cost of a new parking garage with the financing of new parks or street improvements.

With the cost of financing new multilevel parking garages ranging as high as $30,000 per space, many private and municipal owners are realizing that to provide adequate parking is an expensive necessity. Developers now realize that providing the necessary parking for mixed-use projects is a significant portion of their total project costs. Municipalities are becoming more aware that public parking should be wholly or partially self-supporting, instead of being a complete burden to the taxpayers.

This book examines the various methods of financing proposed parking facilities, by both the private and public sectors. Banks, investment banking firms, and venture capital firms provide financing expertise on a client-by-client basis, but a specific summary of parking financing has never before been presented, so as to enable prospective parking project owners to know what to ask for, what to avoid, and what may be possible under current tax laws.

E. Carlton Heeseler

PREFACE

The level of knowledge concerning the financing of parking facilities is high among private parking operators and owners. Internal growth and new projects would not occur, if parking companies were not thoroughly familiar with the ability to finance new projects and acquisitions, at advantageous terms. Understandably, each private parking company jealously guards not only its source of borrowed or equity funds, but also the methodology of the financing used.

The vast majority of new parking projects, however, are planned, financed, and owned by developers, industrial companies, and municipalities. While each of these categories of parking facility owners have knowledge of their respective methods and sources of financing for general corporate or municipal purposes, the level of expertise concerning specifically the financing of parking facilities is sadly lacking. If for no other reason, both private corporations and municipalities frequently include the cost of a new parking project as part of an overall financing package. Developers usually include the cost of associated parking facilities within the overall project development [illegible]

With the cost of building new multilevel parking garages ranging as high as [illegible] analysis of [illegible] parking is an expensive [illegible]

[illegible]

This book examines the various methods of financing proposed parking facilities, by both the private and public sectors. Banks, investment banking firms, and venture capital concerns all finance on expertise on a deal-by-deal basis, but a specific summary of parking financing has never before been presented, so as to enable prospective parking project owners to know what to ask for, what to avoid, and what may be possible under current tax laws.

E. Carlton Heeseler

ACKNOWLEDGMENTS

The author gratefully acknowledges the guidance and assistance of the following persons, without whom this book could not have been written:

Brian M. Carmody, Partner
First Coastal Properties,
Charleston, SC

Robert W. Chamberlin, Senior Vice President
Dean Witter Reynolds, Inc.,
New York, NY

George V. Dragotta, Executive Vice President
National Parking Association,
Washington, DC

Gerard Fernandez, Jr., Partner
Hawkins, Delafield & Wood,
New York, NY

Kim Greene, Assistant Word Processing Supervisor
Shearson Lehman Brothers,
New York, NY

Robert Martin, Vice President
Shearson Lehman Brothers,
New York, NY

Roger C. McDaniel, Senior Vice President
Shearson Lehman Brothers,
New York, NY

Kevin Waymer, Word Processing Supervisor
Shearson Lehman Brothers,
New York, NY

CHAPTER 1

THE PARKING BUSINESS

The parking industry has been characterized as a real estate business with a high tenant turnover. A parking facility must be considered as a building or surface improvement of specialized or limited usefulness. Therein lies the problem, in the basic understanding of the financing of such property in the normal context of corporate finance. Real estate risk increases as the property involved becomes more unique either because of large size or because of a specialized structure to meet the peculiar needs of a particular class of "tenants." To institutional investors the underlying risk in parking facility financing is associated with two factors: (1) the degree to which a parking structure is likely to be able to secure alternative users, and (2) if it is of a type which may find alternative users, the degree to which such alternative demand is variable. If a parking structure is capable of being renovated for alternative users, mortgage financing is more readily available, secured by the assets of the building. The more specialized the structure, the more will any debt financing have to depend on the general credit of the owner.

Most of the time, private developers finance an entire mixed-use development project in one financing package, including any parking requirements which must be met. Developers have already established their sources of capital or they would not remain in business for very long. As a result, the capital cost of included parking facilities is merged into the overall capital costs of the project and the normal requisites of equity and mortgage loan percentages acceptable to the real estate market are carried as a part of the overall financing. Usually, in such cases, the revenues, if any, and the operating and maintenance expenses pertaining to the parking are merely line items in the developer's projections. As a result, lenders analyze the developer's credit on an overall basis, with the parking portion as an ancillary part of the whole picture.

The financing of parking facilities by a private corporation is not a science, but rather one aspect of the parking business. Business is an obligation gradually from funds generated by parking operations rather than be compelled to refinance under uncertain future conditions.

Another external influence which has a significant effect on the cost of alternative financing methods is the impact of Federal income taxes. Corporate income tax rates have put a premium on debt financing as compared with the use of equity, for the reason that debt interest is deductible for tax purposes, while dividends are not. The relative disadvantage of equities is indicated by the fact that new money raised by the sale of stock must earn a premium, equal to the rate of income tax to the company, in order to cover dividend payments, over that which must be earned to meet interest payments on debt. The offsetting factor of no required principal repayment for equity financing is balanced, to some degree, by the higher rate of return demanded by the market for equity participation, as compared to debt obligation interest rates when combined with long term amortization schedules.

STAND-ALONE PARKING PROJECTS

In those instances where a developer is seeking financing for a separate parking project, which may or may not be an adjunct to another real estate project, the analysis of the credit aspects are completely different. Today's lenders will demand more than just a mortgage and pledge of the revenue stream from the financed parking facility. This enhancement to the borrower's credit can take one of several forms:

- a security letter of credit from a bank or other financial institution with an investment grade rating,
- collateral or strong personal guarantees or a strong corporate guarantee of the borrower, or
- pledging as security a leasehold interest, by leasing the facility to a strong corporate user or the host municipality.

The reason why investors want some form of credit enhancement for stand-alone parking project financings is quite simple: virtually without exception, lenders are not convinced that a single, start-up parking project is an economically viable entity. Therefore, they want additional credit enhancement to protect their investment. This same philosophy holds true if the lender is a bank, insurance company, pension fund, or the general investing public, in the case of a public offering of notes.

From a lender's point of view, one of the best forms of enhancing the credit of a prospective borrower and attaining an investment grade rating for new project financing is a letter of credit protecting the security of the loan. Unfortunately, a security letter of credit from an acceptable financial institution is also one of the most expensive forms of credit enhancement, on an annualized basis. In the past several years, the major Japanese banks have become the most aggressive institutions in the field, undercutting the major American banks in the annual rate charged. While still dependent upon the underlying credit of the borrower or the project, Japanese banks in recent months have granted letters of credit for as low as ½ of 1% per year of the outstanding loan balance for borrowers with a good financial history, while money-center American banks are quoting at least 1% per year for the same underlying credit. The typical letter of credit is usually available for a period of time considerably shorter than the 20-to 25-year life of a long term borrowing period and would have to be renewed (usually every five to seven years) to remain in effect.

Frequently, the question is asked: if I can get a letter of credit from a bank, then why do I need separate mortgage financing, i.e., why not draw down on the letter of credit to finance the project? The answer lies in the form of a security letter of credit at relatively low annual rates, as opposed to a line of credit from a bank used to finance construction, at an annual rate of probably not less than prime (currently 9.00%) and undoubtedly variable monthly or quarterly. The letter of credit obtained to secure other permanent financing is only used or drawn upon to prevent a default in mortgage payments due to a lack of current project income. In effect, it is used as a guarantee, not as a source of financing.

Another method of securing project financing is the pledging of collateral or a corporate or personal guarantee. If a borrower has available government securities or similar quality investments which can be deposited in trust to secure outside financing, lenders will finance a new project at reasonably attractive rates. Usually, the lender will require that such collateral be deposited in trust in an amount equal to 1.20 to 1.60 times the amount of the financing to be secured, depending on the type and quality of the securities deposited. Interest on the collateral securities is deposited to the credit of the borrower and can be used to offset the interest cost of the project financing.

While collateralization has been used in the past for project financing, it is rare, primarily because the developer does not have such collateral available. The alternative most utilized by developers is a personal guarantee by the principals or a corporate guarantee by a strong developer company with a good track record of profitable operations over at least a five-year period.

When contemplating personal guarantees, the principals involved should be aware that some restrictions will be placed on their personal finances, including a provision to maintain a specified level of net worth and restricting the sale of personal assets.

Larger parking operating companies usually prefer the alternative of a corporate guarantee as security for a project mortgage loan. Such a loan from a commercial or savings bank or a local insurance company would usually be in an amount equal to 70 to 80% of the total project cost, with the balance representing an equity contribution from the corporation.

Interim financing to pay the cost of construction of a new project may take the form of a line of credit from a local commercial bank. Such financing is short term and can be drawn upon up to 100% of the amount of the line of credit or total project costs, whichever is less. While a corporation will have to pledge its credit and the physical assets of the project during construction, the usual course is to *take-out* the interim construction financing with a longer term mortgage loan or some other form of permanent financing. In this respect, those projects which are constructed with interim money, with no permanent mortgage loan take-out, are at some economic risk due to the high cost of such interim financing and the short term nature of the loan. In effect, the bank granting a short term line of credit for construction would be forced to roll-over the total loan until the developer could find a source of permanent financing.

CHAPTER 2

SOURCES OF FUNDING

The types of investors that purchase taxable equity or debt obligations differ markedly from those that purchase tax exempt bonds. The primary motivation to investors, with respect to the type of investment made, lies in the effect of Federal income taxation on each investor as a class. Different rates of income taxation on different classes of investors result in buying patterns that are fundamentally driven by complicated tax laws. As an example, pension funds and life insurance companies have a lower income tax rate than individuals, banks, or casualty insurance companies. Therefore, pension funds and life insurance companies have little interest in investing in tax exempt bonds in order to gain tax exemption, preferring, instead, the higher yields of taxable obligations.

TAX EXEMPT FUNDING SOURCES

The most recent figures of the Federal Reserve, with respect to the ownership of tax exempt obligations, show the following in 1990:

Type of Investor	*Tax Exempt Bond Ownership in 1990*[1]
Households (individuals)	41.6%
Mutual/Bond funds	12.5
Commercial banks/Trust Companies	14.7
Insurance companies (casualty)	16.3
Money Market Funds	9.4
Other, including corporations	5.5
TOTAL	100.0%

[1]Third Quarter of 1990; see Appendix 5 for similar statistics since 1960 and Appendix 6 for a chart of the holders of long term municipal debt. Appendix 4 shows the holdings of tax free investments, by age group, at the end of 1988.

The enactment of the 1986 Tax Law had a significant effect on the issuance of municipal bonds and on the types of investors who purchase such tax exempt obligations. A recent study shows the changes in buying patterns by investors:

Types of Investors	*Buying Patterns Tax Exempt Investors 1980*	*1989*
Households (individuals)	25.2%	42.4%
Mutual/Bond funds	1.8	19.9
Commercial banks/Trust Companies	42.6	16.3
Insurance companies (casualty)	23.0	16.5
Other, including corporations	7.4	4.9
TOTAL	100.0%	100.0%

The total annual volume of the new issue tax exempt bonds sold in the market reached a peak in 1985 of slightly more than $222 billion. As anticipated, the 1986 Tax Law caused a major decline in new issue volume in succeeding years, to $151.7 billion in 1986 and down again in 1987 to $105.7 billion, the lowest level since 1983, when $87 billion of new tax exempt obligations were sold. Since then, the pent-up demand for financing local and state capital needs caused the total volume of tax exempt bond issues to increase each year, to $115.1 billion in 1987, $114.8 billion in 1988, $125.0 billion in 1989 and $126.4 billion in 1990.

Appendix VII shows the dollar volume and number of long term (over 1 year) issues sold by states and municipalities, including agencies and authorities, for the calendar years 1989 and 1990, by major purpose.

TAXABLE FUNDING SOURCES

According to the statistics reported by the Federal Reserve Bank in December 1990, in their publications *Credit Market Supply of Funds,* the total private domestic holdings of debt instruments and corporate equities, in the third quarter of 1990, amounted to $16,811.7 billion, up from $6,056.4 billion for the same period in 1982, representing a 177.6% increase in seven years. Table 2-1 shows the breakdown of type of debt and equity for the third quarter of 1990. The table in Appendix 2 shows a breakdown of such holdings by major categories of debt and equities for the years 1982 to 1989, first quarters. It is interesting to note that relative proportions of most categories of debt and of equities remained fairly constant or declined slightly over the

Table 2-1. Total Credit Market Debt Outstanding Third Quarter, 1990

TYPE	AMOUNT (1)
U.S. Government securities	$ 3,838.1
Tax exempt obligations	836.3
Corporate and foreign bonds	1,588.8
Mortgages	3,727.1
Consumer credit	794.4
Bank loans	817.6
Open-market paper	611.7
Other loans	828.0
Total debt	$13,042.0
Mutual fund shares	547.3
Other corporate equities	3,222.4
Total debt and equities	$16,811.7

(1)Amounts in billions
Source: Federal Reserve Bank, Credit Market Debt, December 1990

seven years, whereas the major proportional increase occurred in U.S. government securities, from 17.6% of the total holdings in 1982, to 23.1% in 1990. It is also interesting to note that the aggregate debt and equities of corporate business in the United States far outweighs the combined debt of the U.S. government, all state and local governments and residential home mortgages, by about 1⅓rd times.

The same Federal Reserve Bank publication reveals the ownership of various credit instruments. In the table in Appendix 3, selected types of securities are shown by the types of investors. The driving force of Federal income taxation with respect to institutional investors can be seen in the proportionate amounts of tax exempt obligations held by those institutional investors subject to higher rates of income taxes. As an example, commercial banks, mutual funds, and other insurance companies, excluding life insurance companies, hold a total of over 50% of the outstanding tax exempt securities, while this same group of institutional investors only hold less than 14% of the total corporate bonds, equities, and commercial mortgages. Life insurance companies, private pension funds, and state and local government investors, as a group hold 33% of the same class of corporate and commercial obligations, because of their lower or nonexistent income tax burden.

There are no statistics available to show the amount of corporate

qualified bidder. From that point the public offering of the bond issue proceeds in the same manner as a negotiated sale and the successful bidder, whether a bank or an investment banking firm, assumes the same underwriting risk. The issuer then proceeds with the publication of the final official statement or prospectus, showing actual interest rates.

In a competitive sale of tax exempt bonds, the successful bidder, in the vast majority of states, must purchase the obligations at a price not less than par or the face amount of the issue. How is the successful bidder compensated for the purchase and public offering of the bond issue? The bidder will assign an interest rate or a series of rates for serial maturities in the proposal, at a level such that the yield offered to the public at the offering will result in a premium over par to the ultimate investor/purchaser. The successful bidder's overall *scale of yields* will average a return price from the public sufficiently above the par purchase price offered to the municipality, to compensate the firm for sales commission, underwriting risk, and closing expenses. Again, however, changing market conditions could result in the successful bidder having to reduce the offered price to the public, in order to sell all of the bond issue, to an extent that the original built-in spread could be eliminated and a loss sustained by the firm.

DISCLOSURE

Many states have enacted laws designed to prevent the sale of fraudulent securities and to require certain standards of performance on the part of issuers of and dealers in securities, generally referred to as *blue sky laws.* Some require that securities dealers and brokers be licensed, others qualify securities for sale in their states only after pertinent data have been disclosed. A few states rely exclusively on fraud law, wherein action by the state attorney general to prevent the sale of potentially fraudulent securities may be taken. In many states, dealers in securities must supply the applicable state commission or department with required information concerning the proposed offering.

Because of the lack of uniformity of state legislation, the numerous exemptions that were permitted, the sometimes weak administration of existing state regulations, and the fact that the state had no jurisdiction over flotations in interstate commerce, the enactment of Federal legislation many years ago became apparent. The pressure for more far-reaching legislation and control led to the passage of the Federal Securities Act of 1933 and the Securities Exchange Act of 1934 (the

"Act"). The original act of 1933 provided for the administration of the law by the Federal Trade Commission. The Securities Exchange Act of 1934 set up the Securities and Exchange Commission (SEC) and gave administration of both laws to this body.

The Act does not attempt to prevent the offering of speculative securities. The chief duty of the SEC is not to pass upon the investment merits of securities, but to see that investors are informed of the facts concerning securities to be offered for sale. The SEC accomplishes its objectives in several ways:

1. Providing that no securities, other than those specifically exempted, may lawfully be sold or offered for sale in interstate commerce or through the mails unless a registration statement containing complete information is filed with the SEC and becomes effective, and a prospectus in the form required by the Act is made available to the prospective purchaser. Exempted securities include Federal, state, and municipal obligations, securities of common carriers subject to Interstate Commerce Commission (ICC) regulation, banks, insurance companies, and savings and loan association securities offered by nonprofit institutions. The SEC also exempts offerings up to $300,000 from regular registration requirements.
2. Giving the SEC the power to prevent the sale of securities until the proper information, completely and accurately stated, is provided in the registration statement and prospectus.
3. Providing criminal liability and penalties for willful violation of the Act and for willful omissions and untrue statements of material fact in the registration statement and prospectus.
4. Giving the purchasers of securities the right to claim damages from the issuer, principal officers, and directors of the issuing company, accountants, engineers, and others responsible for material in the statement, and the underwriters of the issue, if the prospectus contains an untrue statement of a material fact or omits a material fact.

Under the Trust Indenture Act of 1939, the regulation of securities embodied in the Federal Securities Act of 1933 and the Securities Exchange Act of 1934, was extended to include provisions covering the trust indentures of debt securities. Registration of debt securities is not effective until the indentures conform to the requirements of this act and are qualified by the SEC. The registration statement filed with the SEC and the offering prospectus must include an analysis of the indenture provisions covering such matters as default, authentication by the trustee, and the release or substitution of any mortgage or pledged property.

Private offerings of securities, either debt or equity, are exempted from SEC registration. The SEC has adopted specific regulations governing the maximum number of offerees to whom a new issue may be shown, below which the offering would be considered a private placement.

As mentioned above, state and municipal securities are exempt from the requirement of filing a registration statement with the SEC and receiving an approval for a public offering. However, on June 28, 1989, in Release No. 34-26985, the SEC adopted Rule No. 15c-12 (the "Rule") effective as of January 1, 1990, which introduced certain requirements for underwriters of municipal securities with respect to due diligence and disclosure. In an effort to provide municipal bond investors with a similar level of disclosure for new municipal securities, as is available for corporate securities, the SEC has, by adopting this Rule, put the burden on underwriters instead of the municipal issuers.

The Rule makes it unlawful to act as an underwriter in a primary offering of municipal securities exceeding $1 million unless the underwriter complies with or is exempted from the Rule. Briefly, the Rule requires underwriters to obtain and review an official statement (prospectus) deemed final by the issuer, prior to bidding for or purchasing municipal securities, to contract with issuers to receive sufficient quantities of final official statements within seven business days after any final agreement to purchase and to make them available to any requesting potential customer for a period of up to 90 days, following the end of the underwriting period. Exemptions are provided for municipal securities which are sold in denominations of $100,000 or more, if (1) the sale is limited to no more than 35 investors meeting standards set forth in the Rule, (2) the securities have maturities of nine months or less, or (3) the securities have provisions which allow the investor to redeem or sell to the issuer the securities, at least as frequently as every nine months.

Such an official statement or prospectus must be complete, except for information relating to the offering price, interest rate, selling compensation, aggregate principal amount, principal amount per maturity, delivery dates, any other terms or provisions required by an issuer of such securities to be specified in a competitive bid, ratings and the identity of the underwriters. The Rule states that the issuer makes the determination of whether the official statement should be deemed final, thus eliminating the uncertainty as to how an underwriter could ascertain completeness prior to its review of the document. In its comments with respect to the adoption of the Rule, the SEC indicated that the requirements do not prevent an underwriter from bidding on an issuer's securities in a public competitive sale, even when the underwriter determines that disclosure problems exist, so long as the underwriter receives assurances from the issuer that the disclosure will be corrected and that substantial changes to the document may be made.

The SEC has also updated its interpretations of the responsibilities of underwriters of municipal securities under the general antifraud provisions of the Federal securities laws. It has stated that under these general antifraud provisions, an underwriter must have a reasonable basis for recommending any municipal securities and that by participating in an offering, the underwriter makes an implied representation that it has a reasonable basis for believing the key representations contained in the disclosure documents (official statement). The SEC expects underwriters to review the issuer's disclosure documents in a professional manner for possible inaccuracies and omissions. While not listing all of the specifics in the due diligence procedure, the SEC said that the following factors should be included:

1. the extent to which the underwriter relied upon municipal officials, employees, experts, and other persons whose duties have given them special knowledge of particular facts,
2. the role of the underwriter, i.e., a bidder at competitive sale or purchaser at negotiated sale,
3. the type of bonds being offered,
4. the past familiarity of the underwriter with the issuer, and
5. the length of time to maturity of the bonds.

The SEC has also stated that the presence or absence of credit enhancement (letter of credit or bond insurance) generally would not be a substitute for material disclosure concerning the primary obligor on municipal bonds and thus would not be regarded as a significant factor in assessing an underwriter's due diligence inquiry.

Notwithstanding the obligations of municipal underwriters set forth in the Rule, the SEC emphasized that issuers are primarily responsible for the accuracy and completeness of their official statements. See Appendix 12 for the text of the SEC municipal bond disclosure rule.

CHAPTER 3

THE COST OF MONEY

Using other people's money to finance capital projects cost money. Whether the form of the capital used is debt or equity or whether the cost is called interest or return on equity, the marketplace supplying the capital demands the payment of a charge for the use of their money. The rate of return will vary according to the actual or perceived risk to the investor, which in turn will vary according to the history of the particular industry and the immediate credit standing of the borrower.

Historically, the yield of interest payments on tax exempt municipal bonds has been lower than any other form of investment, when comparing quality and maturity. However, with the advent of the changes in the laws concerning the issuance of tax exempt bonds, the difference or spread between U.S. Treasury obligations and municipal bonds has narrowed. In early 1981, the spread between 30-year Treasury bonds and AAA rated long-term general obligation municipal bonds was over 3.50% (12.54% vs. 9.00%). By June of 1986, however, this spread had been reduced to a slight negative; long term AAA municipals were yielding 0.07% *more* than 30-year Treasuries.

Long term AAA rated corporate (industrial) bonds consistently show higher yields than either Treasuries or municipals of comparable maturity. Since 1981, industrial bonds have shown a spread over Treasuries ranging from a low of about 0.40% to a high of about 1.30% and over municipal bonds in the range of about 0.75% to as high as 3.80%.

As of February 1991, AAA, long-term municipal bonds are yielding about 6.80%, compared to U.S. Treasury 30 year bonds at about 8.20% and AAA, long-term industrials at about 9.25%.

Interest rates or yields on U.S. Treasury obligations are effected by actions of the Federal Reserve Bank in restricting or making money available to the banking system. Such actions of the Federal Reserve are dictated by economic considerations of inflation, recession, or to slow down a booming economy. The market for Treasuries is also effected by the actions and investment attitudes of foreign banks and institutional investors. Basically, all domestic interest rates are governed by the trend movements of Treasury obligations, with the spread between comparable maturities varying somewhat, depending on specific factors that effect a particular type of municipal or corporate class of obligation.

The return on equity demanded by investors, as opposed to a fixed rate of interest on debt financing, is a much more volatile commodity, not necessarily subject to the same influences as interest rates. As an example, yields on medium grade industrial preferred stocks have ranged from 11% to over 13% in the past nine years, whereas common stock yields are significantly below these levels. Generally, however, small, privately placed equity issues will require a yield or return on equity to individual investors of anywhere from 10 to 15%.

COSTS OF ISSUANCE

Regardless of the method of financing a new parking facility, whether by tax exempt bonds, taxable lease obligations, or conventional debt/equity financing, there will be costs of issuance or *soft* costs not related to the *hard* costs of construction or the acquisition of land. Such soft costs can range from as low as ½ of 1% to over 20%, depending on the method of financing. A $5,000,000 general obligation tax-exempt bond issue, sold at competitive bidding, could have legal, printing, and closing costs paid by the issuing municipality, of as little as $25,000. At the opposite extreme, a complex, taxable lease revenue obligation to finance $5,000,000 of hard costs could incur soft costs of close to $1 million, including a debt service reserve; capitalized interest; credit enhancement; and legal, printing, and closing costs.

Current ranges for the various soft costs incurred for tax-exempt and conventional financing are discussed below.

Tax Exempt Bonds

1. The costs of issuance, usually included in the bond issue and paid from bond proceeds include:

   - bond counsel fees, ranging from about $1.00 to $5.00 per $1,000 principal amount of bonds, depending on issue size and complexity

- printing costs of the offering prospectus and the registered bond certificates, ranging from about $1.00 to $3.00 per $1,000, depending on quantity
- bond discount or underwriter's spread, ranging from $5.00 to $30.00 per $1,000, depending on the structure of the financing; i.e., short-term notes vs. long-term bonds, or AAA rating vs. BBB rating; bonds sold at competitive bidding are often required to be sold at par, i.e., no bond discount
- advertising and filing fees are minor, usually amounting to less than 25¢ per $1,000.

2. The interest which must be paid on revenue bonds during a construction period should be included as a cost of overall financing and paid from the proceeds of the bonds; these costs can be calculated based on the plan of financing as follows:

    - if bond anticipation notes are to be issued, the initial amount of the notes could be for the full project costs or for initial expenses, with additional notes issued periodically as required; short-term, one year tax-exempt interest rates currently range from 5.00 to 5.50%; interest on BANs are paid at maturity from the proceeds of the notes; at the completion of construction, when BANs are retired from the proceeds of the permanent financing, the exact amount of interest on the BANs is known and included in the funding
    - if long-term bonds are issued initially for full project costs, the bond interest rate over the construction period will give the amount of capitalized interest, from which should be deducted the estimated investment earnings based on the construction cost draw down schedule; investment grade bond interest rates currently range from about 5.00% for floating rate bonds to about 7.05% for 30-year fixed rate bonds of medium quality.

3. Reserves for the temporary protection of debt service payments or operating expenses may be included in calculating a bond issue and set aside for these purposes from the bond proceeds; the following guidelines can be used to ascertain the amounts required:

    - it is *not* necessary to include either debt service reserves or operating reserves when financing is done by the issuance of general obligation bonds

- when parking revenue bonds or lease revenue bonds are being issued, it is mandatory to include provision in the bond proceeds for the funding of a debt service reserve and, although not mandatory, it is advisable for new facilities to create an operating reserve; a debt service reserve should be one full year's average annual principal and interest but may be not less than the maximum annual interest due on the bond issue; operating reserves vary from two to three months of budgeted operating expenses.

Conventional Financing

Because the typical conventional project financing is approached in a completely different manner than tax-exempt financing, the costs associated with such financing are calculated or estimated differently. The usual procedure for a private parking owner to follow includes obtaining a commitment for a line of credit from a local commercial bank for all of the costs anticipated to the completion of the project. In most instances, however, a commercial bank will *not* grant such a line of credit, unless, (1) the owner has sufficient existing balances in that bank to cover at least some portion of the amount of the line of credit, or (2) the owner has provided the commercial bank with an irrevocable commitment or *take-out* for permanent mortgage financing, plus an agreed owner equity contribution, to take effect upon completion of the parking project, or (3) lacking a *take-out* commitment from the developer, the bank would insist on a first mortgage on the facility, with the option to transform the interim loan to a mortgage loan. Assuming the parking project owner intends to utilize mortgage and equity financing for the project, the following costs associated therewith can be expected:

1. Line of credit costs—these costs can be broken down into two areas:

   - an interest charge on the amount of the construction line of credit drawn down periodically; this may be calculated daily, weekly, or monthly depending on the bank's policies and the negotiated schedule of draw down rights; interest rates for this type of interim financing are seldom less than prime (9.00% currently) and frequently one or two percentage points higher, and

- a stand-by commitment fee may be charged, to recompense the bank for the reservation of the amount of funds committed to the owner's account; this fee is usually in the range of about ½ of 1% on the balance of the commitment remaining after periodic draw downs.

2. Mortgage fees—as a part of the total financing costs, there are two types of fees or charges levied by a mortgage lender which should be included as part of the overall cost of financing:

   - a stand-by commitment fee for the permanent mortgage loan may be charged to recompense the mortgage lender for the reservation of the amount of funds committed to the owner's project upon completion; this fee also is usually in the range of ½ of 1% of the mortgage amount committed, and
   - most mortgage lenders discount the total amount of the mortgage proceeds going to the owner, i.e., *points* are charged upon the take down of the mortgage at the completion of the project; these points range from 2 to 8% of the total amount of the mortgage.

3. Other fees and charges—the developer is subjected to a variety of other expenses, in addition to hard costruction contract costs, plus the developer should receive some reimbursement for the time and effort in putting the project together:

   - legal expenses, i.e., the cost of one or more attorneys who represent the developer in the negotiations with banks, lenders, and the host municipality; such legal expenses can amount to as high as 1% of a project cost, depending on the size and complexity of the project
   - consulting engineering fees for economic feasibility studies, environmental studies, site studies, and functional design of the proposed parking facility (project design work is usually included in the hard costs)
   - the cost of test borings on the site, building permits and other filing costs
   - the developer's fee is usually included in the total project cost, to reimburse the developer for the time and effort (payroll and

out-of-pocket expense) taken to initiate and develop the proposed project; this amount can range from as little as 1% to as high as 10% of total project costs, depending on the length of time it has taken the developer to bring the project to completion and the degree of the developer's equity in the project financing and his negotiated percentage of the net cash flow from the completed project
- all construction projects should have a contingency fund capitalized from the project financing; such a contingency would cover unavoidable work order changes during construction, force majeure and other unforeseen costs; the amount of a contingency fund to be set aside should be not less than 5% of hard construction costs, unless the construction contractor has executed a guaranteed maximum price contract with the owner.

COST OF BORROWING

The cost of borrowing for a developer financing a parking project through a conventional debt/equity package, in addition to the mortgage debt interest rate, is significantly effected by the following factors:

- the ratio of debt to equity
- the amount of the total project costs which can be depreciated by the owner, and
- the term of allowable depreciation

The ability of a parking developer to raise debt and equity capital is dependent upon the attractiveness of the financing package to investors. The debt portion of the package is the easiest to calculate; the interest rate charged by the mortgage lender will be based on the lender's appreciation of the general credit of the borrower, as well as the perceived risks under the terms of any lease agreement, which may be assigned to the lender as security for the mortgage.

The attractiveness of the financing package to equity investors is decidedly more complicated. The key rate which determines that attractiveness, in competition with other investments available to the prospective equity investor, is the after tax return on equity. The formula for expressing After Tax Return On Equity (ATROE), based on current tax laws, is as follows:

$$((EC \times E) + ET \times (((D \div DY) \times ED) - (EC \times E))) \div E = \text{ATROE}$$

where

EC = the predetermined percent of cash flow on equity (rate)
E = the total amount of equity in the financing package (in dollars)
ET = the applicable percentage tax bracket of the equity investor
D = the total amount of fixed assets to be depreciated
DY = the weighted average term of depreciation in years
ED = the percentage of annual depreciation allocated to the equity investors
ATROE = after tax return on equity

As an example, assume the following:

$EC = 9.0\%$.
$E = \$1,250,000$
$ET = 28\%$
$D = \$5,000,000$
$DY = 31.5$ years
$ED = 90\%$

Thus, solving for ATROE shows 11.428% = after tax return on equity.

In this example, equity investors receiving 90% of the depreciation benefits on a project having $5,000,000 of aggregate depreciation on a straight line over 31.5 years, plus an annual cash flow of 9.0%, can expect a total 11.428% return on equity, after application of a 28% income tax bracket.

The interrelationship of the percentage of cash flow, the term of depreciation and the income tax bracket of the equity holder is obvious. The key factor in determining the controllable variables (cash flow percentage, percentage of depreciation allocated to equity investors, and the amount of equity to be contributed) lies in the achievement of an ATROE that is marketable.

Experts in real estate equity offerings indicate that historically, investors demand an ATROE from all sources of 0.50 to 1.50% over medium term Treasury bonds (10 to 15 year maturity). As in other types of investment, this relationship changes from time-to-time depending upon overall changes in interest rates, changes in tax laws, general economic conditions, and the availability of such equity offerings in the market.

MARKET VALUE

A potential lender's approval of a financing package involves the acceptance of the risk involved in comparison with the reward of the return on the investment. A part of the analysis of determining the degree of risk is the assurance of the borrower that the project to be financed will produce an adequate stream of revenues. Another part of that analysis, particularly for privately owned parking facilities, which is frequently overlooked by corporate borrowers, involves the potential market value of the mortgaged parking facility in the event of a possible foreclosure. The definition of "market value," as taken from the American Institute of Real Estate Appraisers publication, *Real Estate Appraisal Terminology,* is:

> "The most probable price in terms of money which a property should bring in a competitive and open market under all conditions requisite to a fair sale, the buyer and seller each acting prudently, knowledgeably, and assuming the price is not affected by undue stimulus."

There are three generally accepted methods for estimating market value. There are also variations within each of the methods depending on the type of property that is being appraised. The following is a brief description of each.

Cost Approach

The Cost Approach is based on the principle of substitution. This principle may be stated as follows: "No one is justified in paying more for a property than that amount by which he can obtain, by purchase of a site and construction of a building, without undue delay, a property of equal desirability and utility."

The Cost Approach consists of four steps:

1. An estimate of land value as if vacant.
2. An estimate of the current cost of reproducing the improvements.
3. An estimate and deduction of depreciation from all causes, if applicable.
4. The addition of the depreciated value of the improvements to the estimated land value.

In the case of prime income producing real estate, the cost of construction plays a minor and relatively insignificant role in valuation. However, the Cost Approach when applied to the evaluation of a proposed building is useful in determining development feasibility when compared to the conclusions of the Market Data and Income Approaches to value. The Cost Approach is considered to be relevant in appraising a special purpose building such as a parking garage.

Market Data Approach

The Market Data Approach to value involves the comparison of recent sales of similar properties in surrounding or competing areas to the subject property. Inherent in this approach is the principle of substitution which states: "when a property is replaceable in the market, its value tends to be set at the cost of acquiring an equally desirable substitute property, assuming no costly delay is encountered in making the substitution." This definition of market value assumes that both seller and buyer are fully informed and neither is under abnormal pressure.

The application of this approach consists of comparing the subject property with similar properties of the same general type which have sold recently or currently are available for sale in the appropriate market area. This comparative process involves judgment as to the similarity of the subject and the comparable sale with respect to many value factors such as location, construction, age, condition, etc. The estimated value through this approach represents the probable price at which the subject property would be sold by a willing seller to a willing buyer as of the date of value.

Income Approach

The Income Approach is based on the theory that the value of a property is equal to the present worth of the net income (cash flow) and reversionary value the property will produce during the remainder of its productive life, over a reasonable holding (ownership) period. There are two separate techniques in the Income Approach which are commonly used: (1) direct capitalization using an overall or equity capitalization rate, and (2) discounted cash flow analysis.

Direct Capitalization. In the direct capitalization method, the net income, or the cash flow after debt service and lender participation, if any, is converted into value utilizing overall or equity capitalization rates, which are abstracted from the market. The pro forma income and expense estimate focuses on the first year of ownership for the property. This technique includes the following five steps:

1. Estimate of effective gross revenue for the first year based on actual traffic counts and estimated economic parking rates in the area.
2. Estimate of property expenses, including real estate taxes and operating expenses for the first year.
3. Deduction of property expenses from gross revenue, to arrive at a first year net operating income estimate.
4. Deduction of debt service and lender equity participation to arrive at a net cash flow.
5. Selection of an appropriate capitalization rate to convert the stabilized net income or the stabilized cash flow into an estimated value.

Most major investors use the Internal Rate of Return (IRR) in analyzing purchases, although some still look only at current cash flow and make no explicit analysis of overall yields over the projected holding period. Typical IRR, which may be used in negotiating the purchase of investment properties, range from a low of 10.0% pre-tax to a high of 15.0%. Most investors feel that an inflation rate of 4 to 5% is applicable for projecting future income and expense increases. Residual values are generally estimated by capitalizing year eleven net income at an 8 to 10% rate.

The overall capitalization rate is dependent to a great extent on the quality of the income stream and the potential for future increases in net income and overall property value at the time of resale. Table 3-1 outlines a summary of yield parameters for institutional investors for various types of investment property. Required overall capitalization rates range from 6 to 10%, depending on a number of factors including property type, location, tenancy, etc.

A parking garage is a fairly unusual type of investment property; it consists of a rental property which has a finite life. An investor should take into consideration the life expectancy (25 to 30 years according to parking consultants) of the property and the implications on the future value of the property as the structure ages. These factors would be reflected in the overall capitalization rate and the IRR. The former assumes a perpetual income stream by definition and is considered to be less reliable as a value indicator than the discounted cash flow technique. The appropriate yield rates for either technique should reflect the nature of the investment, the risk of future capital expenditures, and limited potential future cash flow.

Based on typical transactions and taking into consideration the location, quality, property type, age, and condition of most existing parking structures, a 10% overall capitalization rate should be applicable when capitalizing stabilized net income; a 13% discount rate or IRR, should be appropriate for parking properties when discounting the projected cash flow to the equity and reversion. These rates are somewhat higher than typical rates in the market and are considered to be reasonable for parking property investment.

Discounted Cash Flow Analysis. In the discounted cash flow analysis, the projected net cash flow from the property, plus the reversion, should be converted into an estimate of value utilizing an appropriate discount rate. The discount rate should be estimated, based upon market evidence and recent discussion with sellers, buyers, and their representatives.

A discounted cash flow analysis should be based on the pro forma estimates of future income and expense increases, capital expenditures, and the timing of cash flows over a period of years consistent with the anticipated remaining life of the project. It is the technique most commonly used by purchasers of investment property and is considered to be particularly relevant for parking facilities due to the nature of the potential income stream.

Table 3-1. Typical Internal Rate of Return (IRR)

TYPE OF INVESTOR	PROPERTY TYPE	CAPITALIZATION RATES		IRR	GROWTH RATES	
		INITIAL	REVERSION		INCOME	EXPENSE
Private Investor	Retail Industrial Offices	7.0-8.5% 8.5-10% 7.5%-9.5%	100 basis points higher	10.5-12%	5%	5%
Pension Fund	Retail/Ind. Apts. Offices	7.5-9.5%	9-10%	11-13%	5%	5%
Pension Fund	Retail R&D/Industrial Apts. Officcs	7.5-9.5%	9-10%	11-13% unleveraged 13.5-15% leveraged	5%	5%
Pension Fund	Reg. shopping centers	7-9%	8-9.5%	10.5-12%	5%	5%
Insurance	Offices Hotels Retail	6.5% 6.5% 8.0-8.5%	N/A	10.0-11.5%	5%	5%
Life Insurance	Fully Leased properties of all types	7-10%	10-12% depending upon property type	10.5%-13%	1% below long term inflation factors	tracking
Private Investor	Offices Malls	6-7%	7.5-8.5%	11-12%	4.5%	4.5%
Bank	Offices	8-9%	8.5-10%	6.5-7.5% above growth rates	4-5%	4-5%
Pension Fund	Offices R&D/Industrial Retail	8-10%	9-11%	12%	4-5%	4-5%
Real Estate Investor	Malls Offices Industrial	6.5-7% 8.5-9.5% 8.5-9.5%	8.5% 8.5-9.5% 8.5%-9.5%	10.5% 11.5-12.5% 11.5-12.5%	4-5%	4-5%
Pension Fund	Reg. shopping centers Class A offices Industrial	7-8% 7-8% 9%	8% 8% 9.5%	11% 10.5% 12%	5%	5%

NOTE: N/A = not applicable.

CHAPTER 4

1986 TAX LAW

The 1986 Tax Law and subsequent amendments had a major effect on the attractiveness of alternate methods of financing parking facilities, including public and private projects. It removed parking as an exempt activity for which tax exempt Industrial Development Bonds (IDB) could be issued, put severe restrictions on what constitutes a public purpose for municipalities and removed many of the tax benefits previously available to real estate investors. The current criteria in the Tax Law effecting tax exempt financing and taxable financing are briefly descibed below.

TAX EXEMPT FINANCING

The 1986 Tax Law made the method of financing a public parking facility a decision which now must be made early. The clear lines of distinction between what may be financed with tax-exempt bonds and what has to be done with taxable financing are set forth in this Law. Basically, the tax-exempt criteria are as follows.

Public Use

Not less than 90% of the available spaces in the financed project must be available to the general public, on a daily, monthly, or yearly basis, exclusive of governmental or nonprofit institutional users.

Use of Bond Proceeds

Not less than 95% of the total proceeds of the bond financing must be spent solely for the construction of the public parking spaces, including soft costs related thereto.

Corporate Guarantees

Not more than 10% of the annual debt service of the bond financing may be paid for or guaranteed by a corporate or nonpublic entity on a long-term contractual basis.

Management Agreements

Any management agreement for the operation of the parking facility may not exceed five years duration, must provide for either a periodic flat fee or a fixed percentage of gross revenues, and must give the owner of the facility the option to cancel at the end of any two year period.

Public/Private Partnerships

Early planning, with respect to the use of a proposed parking project and which party will assume the financial liability for the debt to be raised, will determine the basic method of financing, i.e., tax-exempt or taxable. A public/private partnership in the development of new parking facilities may be done with either tax-exempt or taxable financing, depending on the final decisions of the private developer and the municipality with respect to the following:

1. the operating profit motive of the developer vs. the public purpose motive of the municipality;
2. the economic feasibility of the project, as it would effect the profit motive of the developer; and
3. the plans of the municipality concerning long-term leasing of parking spaces or the inclusion of retail commercial space in the parking facility.

A public/private partnership in the development of new parking facilities can be initiated by either a private developer or by the municipality. In effect, it could be an office building developer looking for the municipality to finance and assume the liability of a parking facility, which local zoning regulations dictate must be provided or conversely, it could be a municipality looking for a developer to build and finance a downtown parking facility to alleviate traffic and parking congestion. Depending on the circumstances, each partner may have to accept some responsibility and liability for the financing involved.

GENERAL TAX EXEMPT CRITERIA

The criteria for tax exempt financing described above is a somewhat simplistic interpretation of current Federal tax law, as it applies specifically to the financing of parking facilities. The major changes wrought by the 1986 Tax Law did not single out parking financing but caused the tightening of the rules for distinguishing between traditional governmental bonds and bonds with more than an insubstantial amount of "private" use or involvement—private activity bonds. Basically, Code section 141(a) provides that a bond is a private activity bond if it meets either the "private business test" of Code section 141(b) or the "private loan financing test" of Code section 141(c) (Internal Revenue Code of 1986, as amended). The 1986 Law reduced the threshold for determining whether a bond meets the private business test to 10% (as opposed to 25% under the pre-1986 law). Thus, issuers of governmental bonds that are not intended to be classified as private activity bonds must be more careful in limiting private involvement. As under pre-1986 law, however, a bond meets the private business test only if:

- more than 5% of the bond proceeds are used to finance the capital expenses of a trade or business carried on by any person other than a governmental unit ("private use test"), *and*
- more than10% of the debt service on the bond is repaid from or secured by revenues from a private trade or business ("security test").

For these purposes, use of a parking facility by a member of the general public is disregarded (e.g. a public parking facility may be financed with governmental bonds, even though it is used by private individuals, as long as it is open to the general public; Code section 141(b)(6)).

The second way that a purported governmental bond would be treated as a private activity bond is if the bond meets the private loan financing test in Code section 141(c). Basically, this test is met if more than 5% (or $5,000,000, if less) of bond proceeds are used (directly or indirectly) to make or finance loans to persons other than governmental units. For purposes of the private loan financing test, a "loan" may arise from a direct lending of bond proceeds or may arise from transactions in which the indirect benefits that are the economic equivalent of a loan, are conveyed. There is an exception to the private loan financing restrictions, under Code section 141(c)(2), for "deemed loans" arising from certain tax assessment bonds. For example, issuers of water or sewer bonds typically assess taxes on surrounding users of the facilities to meet debt service. Absent Code section

141(c)(2), the tax assessment structure could be characterized as a loan to the users of the water or sewer facility, including major industrial or utility users.

If a bond is classified as a private activity bond, in order for the interest to be tax exempt the bond must be a "qualified" private activity bond (Code section 141(e)). Under the 1986 Law rules, only the following types of private activity bonds are qualified:

1. Exempt facility bonds, i.e., a bond where at least 95% of the proceeds are used to provide:
   - airports
   - docks and wharves
   - mass commuting facilities
   - some residential rental projects
   - local heating or cooling facilities
   - sewage facilities
   - solid waste disposal projects
   - water facilities
   - electric or gas facilities
   - hazardous waste projects
2. qualified mortgage bonds,
3. qualified veterans' mortgage bonds,
4. qualified small issue bonds (basically limited to manufacturing bonds and qualified refundings),
5. qualified student loan bonds,
6. qualified redevelopment bonds, and
7. qualified section 501(c)(3) bonds, i.e., bonds issued by or on behalf of not-for-profit entities.

In addition to limiting the scope of governmental bonds and the types of private activity bonds that may be issued as tax exempt, the 1986 Law included various overall restrictions on the issuance of such bonds. For example, most private activity bonds issued under the 1986 Law are subject to the state volume cap limitation. Code section 146 sets the amount of the volume cap for each state equal to the greater of $50 for every individual who is a resident of the state or $150 million, with qualified section 501(c)(3) not-for-profit entity bonds being the most significant exception. Secondly, under Code section 147(q), no private activity bond may be qualified for tax exemption if the issuance costs (bond counsel's fees and underwriter's spread), in an amount in excess of 2% of the proceeds of the issue, are financed from bond proceeds. Third, Code section 148(d)(2) provides generally with respect to all tax exempt bonds, that no more than 10% of bond proceeds many deposited in a debt service reserve fund, even if the money in such a reserve fund is restricted in the investment of funds to the average yield of the tax exempt bond issue for which it is a reserve.

THE FUTURE FOR TAX EXEMPT FINANCING

The future for tax exempt financing is, at present, disappointing and unattractive. In 1988, the U.S. Supreme Court held that tax exemption of the interest on municipal bonds is not protected by the Constitution and is, therefore, subject to the whim of the Congress. Since 1984, the Congress has been whittling away at the ability of states and municipalities to issue tax exempt bonds, by placing various restrictions upon those issues, which are entitled to tax exemption, the latest being embodied in the far reaching 1986 Law. The basic attack upon municipal financing as to what should be tax exempt and what should not, is in the concept and definition of what constitutes a public purpose. So far, the Congress has said that pubic purpose financing can be tax exempt. But that which is a public purpose, in the minds of Congress, is highly subject to interpretation. Such things as roads, bridges, sewer systems, and public parking garages are still considered public purposes. But for how long? At what point in the process of deficit elimination and raising Federal tax revenues will such public purpose projects be declared otherwise? All it takes is an act of Congress. Over the long term, the continued ability of municipalities to issue tax exempt bonds is in serious jeopardy, unless current, slow efforts to achieve a Constitutional amendment are successful.

REBATE REGULATIONS

The 1986 Tax Law (Section 148, Internal Revenue Code of 1986) set forth the basis for detailed and complex rules with respect to the prohibition of any arbitrage earnings associated with the issuance of tax exempt bonds. The concept of these arbitrage regulations is that the earnings from the investment of construction funds, reserve funds, escrow funds, or other proceeds obtained from the sale of tax exempt bonds may not exceed the "yield" of the bond issue from which the proceeds originated. Any earnings from investments in excess of the calculated yield are required to be rebated to the U.S. Treasury.

Failure to rebate such excess earnings or arbitrage could cause the IRS to declare the interest on the original bond issue subject to Federal income taxes, possibly retroactively to the original date of issuance. For issuers of tax exempt bonds, the key to the calculation of a rebate, if any, is contained in the IRS regulations defining the method of calculating the "yield" of a bond issue. It is defined as being the discount rate that, when used in computing the present value of all of the payments of principal and interest and all the fees payable for a "qualified guarantee" with respect to the bond to the calculation date, produces an amount equal to the present value of the issue price of the bond. Fees paid with respect to letters of credit and other credit support facilities, such as municipal bond insurance, that constitute qualified guarantees are taken into account in computing the yield on the bond issue.

Such arbitrage calculations can be made by issuers of fixed rate tax exempt bonds, immediately after the closing of the financing. However, issuers of variable rate tax exempt obligations must go through these calculations each time the interest rate changes. In either instance, however, it is not necessarily the fact that a rebate may be owed or paid to the Treasury, which causes tax exempt issuers a concern, but the possibility of the omission of the calculation or an incorrect calculation which might lead to the issue being declared taxable. For small, relatively unsophisticated municipalities, the arbitrage regulations have become a matter of great concern.

Late in 1989, Congress passed an amendment to the 1986 Tax Law, permitting some relief from the onerous arbitrage and rebate provisions in the original law. Under the new amendment, issuers of tax exempt governmental and not-for-profit 501(c)(3) obligations are exempt from the arbitrage rebate provisions, if the issuer spends 10% of the bond proceeds within six months after the date of issuance of the bonds, 45% by the end of the first year, 75% within 18 months, 95% within two years, and 100% within three years. The amendment allows an issuer of tax exempt bonds to elect, at the time of issuance, one of two penalties, if the spending schedule was not met: either rebating the arbitrage profits or paying a penalty equal to 1.5% of the amount of bond proceeds not expended by the scheduled date.

TAXABLE PRIVATE FINANCING

The primary effects of the 1986 Tax Law on the private financing of parking garages are discussed.

Income Tax Rates

The 1986 Tax Law states that after December 31, 1987, there would be no difference in the tax rate on ordinary income and capital gains for individual tax payers. As a result, there is no longer any benefit, in terms of tax rate, to have capital gains from a real estate investment rather than ordinary income. Subsequent efforts to reinstate a separate capital gains tax have been unsuccessful.

Passive Loss Provisions

Generally, the 1986 Tax Law disallows both (1) the deduction of passive investment losses against any kind of non-passive investment income (no deduction is allowed against any other income of the tax payer) and portfolio income and, (2) the use of credits from passive investments against taxes, other than those allocable to passive investments. Passive investments are defined as those for a trade or busines activity in which the taxpayer does not materially participate and all rental activities. The passive loss provisions generally apply to individuals, estates, trusts, and personal service corporations.

Alternative Minimum Tax (AMT)

The 1986 Tax Law added a new alternative minimum tax for corporations at 20% and expanded the alternative minimum tax (AMT) for individuals to 21%. Under the 1986 Tax Law, if the AMT income is greater than 1⅓ times the regular taxable income, an individual tax payer will be subject to AMT. In addition, however, the scope of the AMT has been broadened, with respect to the preferences and adjustments which are permitted. Some of the more significant changes in the 1986 Tax Law, which effect real estate transactions and investments in real estate limited partnerships, include: (1) the depreciation on real property being calculated on a straight-line basis over 40 years, with varying terms for other types of property and equipment, (2) the prohibition of the use of the installment method for the disposition of rental property where the purchase price exceeds $150,000 (thus the taxpayer must recognize the entire realized gain on the sale in the year of sale for purposes of calculating the AMT) and, (3) the new passive loss provisions mentioned also apply in computing the AMT.

Depreciation

The 1986 Tax Law retains Accelerated Cost Recovery System (ACRS) but with substantial modifications. First, residential real property must be depreciated on a straight-line basis over 27½ years and nonresidential real property must be depreciated on a straight-line basis over 31½ years. A taxpayer may elect irrevocably to apply an alternative depreciation system using a straight-line method over 40 years, in which case, the depreciation for regular income tax and the alternative minimum tax would be the same. Second, a lessee is now required to recover the cost of lessee leasehold improvements, over the ACRS recovery period of the improvement, rather than over the shorter period of the real property's basic ACRS recovery period *or* the period remaining on the lease. The changes in ACRS depreciation made by the 1986 Tax Law in general provide more favorable depreciation allowances for certain personal property but less favorable allowances for real property.

CHAPTER 5

SECURITY FOR PARKING FINANCING

Debt financing for a parking facility, whether privately or publicly owned, requires sufficient legal security to ensure the investor that the debt will be paid, along with the agreed interest rate, and that the parking project or the parking system of which the project may be a part, will be operated in a prudent and efficient manner. Any mortgage indenture, trust indenture, or bond resolution creating that parking debt must provide for the allocation of the revenue or income of the debtor company or municipality and must provide for the protection of the debt holder and the remedies available to the debt holder if an event of default occurs. Whether the debt obligation is payable solely from project income or is a general corporate or municipal credit, the lien must be clearly defined. Failure to include a clearly defined source of income and lien thereon to the benefit of the debt holder will result in an obligation which would be unmarketable, at public offering or private placement.

PROVIDING FOR BOND SECURITY

Every municipal bond must be authorized by formal action of the issuer, in accordance with applicable state laws and, in some cases, local charter provisions. This formal action may take the form of a simple resolution of the governing body of the municipality authorizing the issuance of general obligation bonds. In such a case, a recitation of the pledge of the municipality to make the required debt service payments from any available funds and to levy unlimited ad valorem property taxes for such payments, is sufficient to provide for the security behind the bond issue. The authorizing bond resolution then becomes the contract between the bondholder and the municipality, during the life of the bonds.

When revenue bonds are issued by a municipality, to finance a parking facility or any other public purpose, the authorizing documentation executed by the governing body is significantly more complicated. By definition, a revenue bond is payable from an anticipated stream of revenues from one or more enterprises owned and usually operated by the issuing municipality. A revenue bond may also include a lease rental revenue bond, under which the municipality becomes the ultimate owner at the end of the lease term. For purposes of this discussion, it is assumed that the municipality has *not* pledged its full faith and credit to a guarantee of such revenue bonds.

The basic documentation securing a municipal revenue bond can take one of two fundamental forms:

1. a bond resolution
2. a trust indenture

Both forms have similar features with respect to the security of the bond issue. A bond resolution constitutes a contract directly between the issuer and the bondholder and sets forth all of the terms and conditions pursuant to which the bonds are issued and the pledges and covenants of the municipality concerning the operation of the enterprise and the source and payment of revenues therefrom. A trustee or paying agent bank is named in the bond resolution, primarily for the purpose of providing a means of depositing revenues, holding certain funds for the issuer in trust and paying bondholder's interest and principal when due. In the case of a bond resolution, the trustee or paying agent bank is acting as a fiduciary for the issuer and does not necessarily represent the bondholders. Pursuant to a bond resolution, any bondholder has the right of suit against the issuer, in the event of a default by the issuer of any terms of the bond resolution.

A trust indenture is a document executed between a bank trustee and the issuer, pursuant to which the issuer authorizes the bonds to be issued. It contains the same pledges and covenants of the municipality concerning the operation of the enterprise and the source and payment of the revenues as a bond resolution. In this instance, however, the pledges and covenants are made through the trustee to the bondholders, whereby the trustee assumes the obligation for and on behalf of the bondholders to monitor the terms and conditions of that contract with the bondholders. When a trust indenture is the authorizing document, a bondholder is specifically denied the right to sue the issuer in the event of a default, the trustee bank assuming that right of suit for and on behalf of all of the bondholders.

The trust indenture form of bond security is prevalent for authorities and special agencies, as opposed to states, counties, and municipalities, most often because it is required by state laws. In addition, however, the trust indenture provides a degree of fiduciary professionalism for the bondholder, not present in a bond resolution.

The terms and conditions present in a trust indenture, and for the most part included in a bond resolution for revenue bonds, should include the following broad categories:

1. definitions of terms used, including the bond form;
2. authorization for the issuance of the bonds, including the specific terms thereof, such as dates of interest payment, maturity dates and rates of interest;
3. orders concerning the disposition of the proceeds from the sale of the bonds;
4. authorization for the issuance of additional bonds payable from the same source of revenues and the conditions thereof;
5. covenants of the issuer concerning the construction of the project being financed with the bonds;
6. creation of a construction fund, into which bond proceeds are deposited, and controlling the investment and payments therefrom;
7. covenants of the issuer with respect to the rates charged by the enterprise or the facility being financed;
8. creation of the various revenue, debt service, and reserve funds through which the revenues from the enterprise flow (flow of funds);
9. assignment to the trustee of all revenues from the enterprise;
10. security for the deposits made to the trustee and the investment of the various funds;
11. provisions governing the optional and, under certain conditions, the mandatory redemption of bonds prior to the stated maturities;
12. other covenants of the issuer respecting the operation of the enterprise, insurance requirements, use of the facilities, sale or lease of the facility, and any equipment therein, accounting and audit procedures and consulting engineering reviews;
13. provisions concerning what constitutes a default by the issuer and remedies available to the trustee in such event;
14. provisions concerning the trustee's duties and responsibilities and concerning the appointment of successor trustees; and
15. provisions governing the evidence of bond ownership, immunities for personal liabilities of the issuer, defeasance of out-

standing bonds and amendments and modifications to the trust indenture.

Of primary interest to bondholders is the section of a trust indenture dealing with the collection and distribution of the revenue stream from the facility or enterprise financed with the proceeds of the bond issue. A typical trust indenture or bond resolution should show a "flow of funds" as follows:

1. all of the revenues and other income from the facility or the enterprise system that are pledged to the payment of the bonds should be deposited into a designated revenue fund, which fund is subject to a lien and charge in favor of the bondholders,
2. the first and prior order of disbursement from the revenue fund should be for the payment of direct and normal operating expenses, including maintenance and repairs, of the facility or enterprise system, as determined by an annual budget and/or generally accepted accounting principles,
3. the next order of disbursement from the revenue fund should be for current debt service requirements, interest, principal and sinking fund, for the ensuing fiscal year,
4. subsequently, any moneys remaining in the revenue fund should be disbursed as follows:

   - to make up any deficiencies in debt service or operating reserve accounts,
   - to accumulate in an improvement or renewals and replacement fund an amount which may be recommended by the consulting engineers,
   - to accumulate a capital improvement fund, if further expansion of the enterprise system is contemplated,
   - a redemption fund for the optional retirement of any outstanding bonds subject to redemption, prior to the stated maturity, and
   - any surplus moneys available, not needed by the facility or the enterprise system as a whole, can be repaid to the general fund of the municipality.

Usually, a trust indenture or bond resolution will include a covenant by the issuer to charge sufficient rates at the facilities involved, such that the aggregate revenues will cover all of the required payments mentioned above, at least through the accumulation of a reasonable capital improvement reserve. Many times, this is accomplished by requiring the rates to be sufficient to "cover" current debt service requirements (in the order described above) by a factor of from 1.25 to 1.50, with the coverage above debt service requirements being sufficient to satisfy the ensuing flow of funds requirements.

Another primary covenant of interest to bondholder's security is the provision in a trust indenture or bond resolution relating to the issuance of additional bonds payable from the same source of enterprise revenues pledged to outstanding bonds and the test of earnings vs. debt service which must be met. Additional bonds issued for the purpose of refunding outstanding bonds, in whole or in part, should be permitted, with suitable restrictions to assure bondholders that such a refunding will accomplish the avoidance of a default situation or result in a lower aggregate level of annual debt service payments. Such refunding obligations do not cause any dilution of earnings coverage available to bondholders and, in certain instances of reduced interest rate levels, can be advantageous. Additional bonding for improvements or new facilities within the enterprise system should be permitted, but within very specific limitations. Unlimited issuance of additional parity bonds puts the bondholder at the mercy of the management of the enterprise system or the governing body of the issuer, with respect to the potential dilution of the limited earnings power of the enterprise to cover debt service on outstanding bonds. It is usual to see an *earnings test* included as a limitation on additional bonds. The terms of such an earnings test can vary from very conservative to very liberal.

As an example, a conservative earnings test to be met prior to the authorization of additional parity bonds might include the following:

> The net pledged revenues available for current debt service in each of the preceding three fiscal years must be equal to 1.35 times the maximum annual debt service over the life of all outstanding bonds payable from such revenues, plus the maximum annual debt service estimated for the proposed additional bonds, with no provision made for the inclusion of any estimated revenues from the new facility to be financed with the additional bonds.

A liberal earnings test to be met prior to the authorization of additional parity bonds might be as follows:

> The net pledged revenues available for current debt service in the most recent fiscal year, when added to the estimated average annual net revenues available for debt service over the first five years of operation of the new facility to be financed with the additional bonds, must equal 1.20 times the aggregate average annual debt service of outstanding bonds, plus the additional bonds to be issued.

Obviously, the liberal earnings test would permit the issuer to issue additional parity bonds to finance a project with a projected negative earnings potential, in effect drawing a subsidy from established system projects and diluting the bondholders protection from earnings. Many variations of this basic earnings test have been devised by issuers and bond counsel. However, it is safe to say that the more liberal the earnings test for additional bonds, the lower the investment grade of the bonds, all other things being equal.

All of the provisions of a trust indenture or bond resolution are drawn for the protection of the bondholder and to provide a vehicle within which the issuer, as operator of the enterprise financed, can reasonably and prudently operate and maintain the project. Because of the impact of the 1986 Tax Law restricting the use of tax exempt financing, an additional segment in new trust indentures and bond resolutions has been addressed. The arbitrage restrictions in the 1986 Tax Law have necessitated careful attention to the investment of funds created in a trust indenture, particularly those funds received from the sale of the bond issue. Specific recitations must now be included to the effect that the issuer will abide by the provisions of Federal tax law with respect to the investment of all funds, such that the bonds will not be deemed *arbitrage bonds,* i.e., that the investment earnings will not exceed the limitations in the tax code. In addition, new trust indentures include the creation of a *rebate fund,* specifically to funnel any unanticipated investment earnings thereto, for later rebate to the Internal Revenue Service.

In order to provide an additional measure of flexibility to finance capital improvements to an enterprise system, municipal issuers will sometimes include in the original trust indenture or bond resolution a provision permitting the issuance of *junior* lien obligations. By including a fairly liberal earnings test for the issuance of such additional bonds at a point in the flow of funds *after* the debt service requirements of outstanding bonds, the issuer will protect current bondholders from any dilution of earnings. If the level of earnings from the enterprise is sufficient to support such a junior lien obligation, the issuer can proceed to raise additional capital funds from this source.

MORTGAGE OF THE FACILITY

Frequently, but by no means prevalent, a mortgage on the facility being financed or on the enterprise system securing a bond issue, will be granted in favor of the trustee or the bondholders, as an additional provision of a trust indenture or bond resolution. For municipal obligations, the inclusion of a mortgage is dependent upon state law; some states prohibit the ability of a municipality to grant such a mortgage on municipally owned property.

The existence of a mortgage provides some additional measure of security to a holder of a municipal revenue bond. The possibility of a foreclosure in the event of a continuing default on the part of the municipal owner and the loss of the property and equipment to bondholders or a trustee, is a significant deterrent to the municipal owner. From the practical side, however, holders of municipal bonds do not want to end up with a parking garage or a sewer system in the event of a default. Basically, they would prefer to see the municipal owner improve the revenue generation of the facility or meet the debt service obligation from other sources of revenue available to the municipality, whether or not such revenues were legally pledged to the bonds.

In all instances of revenue bond financing, whether through the structure of a bond resolution or a trust indenture, a perfected security interest in the pledged property and equipment should be granted to the trustee or the bondholders, in accordance with uniform commercial codes in all states.

PROVISIONS IN A LEASE AGREEMENT

Municipal lease rental revenue bonds or certificates of participation are secured, in addition to the authorizing trust indenture or bond resolution, by the provisions of the lease or lease purchase agreement between a lessor and a municipal lessee. For this type of financing, the bond holder is one step removed from having a lien on project revenues, as is the case with direct revenue bonds. As a result, the basic security offered to the bond holder in a lease or lease purchase financing is embodied in the provisions of the agreement between the lessor and the lessee, as assigned to the trustee under a trust indenture or to the bond holders directly under a bond resolution. While the lessor of a parking project could be either a private corporation or a public entity, it is assumed that the lessee would be a bona fide political subdivision or agency thereof, in order to achieve tax exemption for the capital financing involved.

The terms and conditions of a lease or lease purchase agreement, used as security for lease rental financing, should include the following broad categories:

1. a recital of the intent of the parties to the agreement and definitions of the terms used therein,
2. representations and warranties of the parties, with respect to their power to enter into such a lease,
3. lease of the project, term and rent payable, which may or may not be dependent upon the completion of the project,
4. provisions for the termination of the lease,
5. provisions for the operation, maintenance, use and inspection of the facilities leased,
6. restrictions concerning alternations, improvements, demolitions and removals of equipment,
7. disclaimer of the lessor with respect to manufacturer's warranties,
8. provisions respecting damage and restoration of the project and the application of insurance proceeds,
9. provisions respecting the possible condemnation of the project and the application of condemnation proceeds,
10. restrictions concerning the assignment of the lease agreement, subleasing and indemnification of the lessor,
11. a recitation of those events which would cause a default under the agreement and the remedies of both parties,
12. provisions for the optional purchase of the leased facilities by the lessee, prior to the intended term of the agreement,
13. miscellaneous provisions concerning the obligations of both parties, the recognition of the security interest of bondholders, accounting and auditing requirements, notices and arbitration procedures.

That portion of a lease agreement of primary concern to bondholders, is set forth in those sections dealing with the term of the agreement and the rental payable thereunder.

Obviously, the term of the lease agreement must not be less than the term of the financing, whether lease rental revenue bonds issued by a municipality or certificates of participation issued by a trustee.

The term of the agreement is usually stated as either a definite date which is the final maturity date of the bonds issued or as an indeterminate date when all outstanding obligations secured by the lease agreement have been retired and paid in full.

The provisions of the lease agreement concerning the payment of lease rental should be stated in two basic parts:

1. the obligation of the lessee to make a specific rental payment equal to the current requirements of interest and principal or sinking fund, usually five days prior to the due dates thereof, on any obligations issued to finance the leased project, and
2. the obligation of the lessee to operate and maintain the leased facility and to pay all costs and expenses in connection therewith.

The obligation of the lessee to make the payments required under the lease agreement may be restricted to the revenues available from the leased facilities or may be a "general obligation" of the lessee, i.e., payable from whatever sources of revenue available to the lessee. Obviously, bond holders prefer the latter form of security, which may remove the economic vulnerability of the parking project from consideration.

When the obligation of the lessee to make the required payments pursuant to a lease agreement is considered to be a "general obligation" or payable from any funds available to the lessee, such an obligation may be either continuous over the term of the lease or may be subject to annual appropriations by the governing body of the lessee municipality.

While either form of the lessee's obligation could be present in lease rental revenue bonds or certificates of participation, it is usual for certificates of participation to be secured by a lease rental payment subject to annual appropriations. The restrictions of state law governing the ability of municipalities to enter into lease agreements determines the ability of the lessee municipality to enter into a continuous lease rental payment obligation. Most states require that such a long term, continuous lease payment obligation be considered a general obligation debt of the municipality. Such a requirement reduces the advantages to the municipality of the leasing method of financing. See Appendix 13 for a schedule of municipal leasing in all 50 states.

Even though a lease payment is subject to annual appropriation, the resulting financing can be marketed, given adequate project essentiality and community support. The resulting lease obligation then has the advantage of not being considered as a general obligation debt of the municipality under state statutes. The reason why most certificate of participation methods of financing parking projects can accept the lease rental payment subject to annual appropriation, is that the lessor is usually a private entity, sometimes a developer or parking company, which, in the event of non-appropriation and termination of the lease, could retake ownership and operate the parking facility, making debt service payments to the trustee directly.

Another provision of the lease agreement, of particular interest to bondholders, is the date of commencement of lease rental payments for debt service. The commencement of the lessee's obligation to operate and maintain the parking facility begins when the project is substantially completed and available for occupancy. However, the obligation to make rental payments for debt service should begin on that date when funds in the capitalized interest fund are exhausted. Normally, in such financing, interest on the bonds or certificates is paid from set-aside bond proceeds to a specific date estimated to cover the construction period. If for any reason, the construction period is delayed beyond this date, the lessee should be obligated to begin the lease rental payments, regardless of the status of completion of the project. In addition, the obligation of the lessee to make lease payments must be stated in the lease agreement to be an obligation not subject to abatement, reduction, abrogation, waiver, diminishment, or modification for any reason; including damage, destruction, or loss of the leased facility.

The lessor of a parking facility leased to a municipality may or may not be required to assume the obligation of the construction and completion of the facility being financed. In many instances, the lessee municipality desires to control the design and construction process, drawing down from the bond construction fund on invoices for work done. In such a case, the lessor may designate the lessee as agent of the lessor to execute the construction contracts and supervise the construction and payments, for and on behalf of the lessor. This arrangement does not, however, absolve the lessor from the basic obligation of construction and completion of the project, within the overall terms of the lease agreement.

Similarly, the lessee municipality may sublease the project or execute a management contract for the operation of the project, to a private or public entity. The terms of such a sublease or management contract may mirror the lessee municipality's operating obligations in the lease agreement but will not absolve the municipality's basic obligation to operate and maintain the leased property. If the financing for the parking project is exempt from Federal income taxes, care should be taken with respect to any subleasing arrangements or any management contracts, that the provisions of the 1986 Tax Law are not violated such as would jeopardize the tax exempt status of the financing.

Non-Appropriation Of Lease Payments

In assessing the cash flow support of instruments backed by a lease or lease-purchase agreement, investors and rating agencies evaluate the possibility of an event of non-appropriation by the lessee. In this regard, the nature of the facility being financed and its importance to the lessee's role in providing essential services are pivotal. Lease purchase financing of nonessential public facilities, such as convention centers, is not as well received by investors as such financing associated with more vital facilities. A parking facility located in the downtown area of a city, backed by strong parking enforcement procedures, would represent a category of essential use facilities. Investors, for this reason, regard lease-purchase financing involving such facilities as representing a strong and attractive credit. To formalize this situation, investors sometimes request the state or local government lessee to provide a letter setting forth the essential nature of the leased property.

Investors traditionally have been comforted by the inclusion of a nonsubstitution clause in the lease-purchase agreement, which would preclude the governmental body from acquiring the same or a similar facility for a specific period of time in the event of non-appropriation. This can reduce the risk of early termination of the lease for other than the legitimate absence of sufficient funds. Typical nonsubstitution language would read as follows:

> "The lessee agrees that in the event the lessee exercises its right to terminate this lease in accordance with the non-appropriation provisions contained herein, the lessee will not purchase, lease, or rent other property for the purpose of performing the functions and

projects which were to be performed by the leased facility for a period of one year from the date of termination of this lease."

The rating agencies generally rate lease-purchase issues subject to annual appropriation at least one full grade below the general obligation rating of the lessee. For example, if a state or local government is rated "AA," the lease-purchase issue would generally be rated "A." This rating policy stems from the risk of non-appropriation. Factors which are positive in the establishment of a rating include: (1) extremely strong security features in the lease including a lengthy lease term, automatic lease renewals, a contractual requirement to make lease payments or a nonsubstitution clause; (2) additional security backing (such as insurance or a bank letter of credit) or specific additional revenues pledged toward payment of the bonds, (3) powerful evidence of the essentiality of the facility; and (4) clear statutory authority for leasing and historical experience by the lessee in making appropriations for tax-exempt municipal leasing. A lower rating will result in a higher interest rate as compared to the lessee's general obligation bonds. If the state or local government has the choice of using either a general obligation bond issue or a lease-purchase financing for a specific project with equal annual payments, either financing mode will have a similar impact on the governmental body's overall credit rating.

A municipal bond insurance company's primary underwriting concern is also the risk of non-appropriation. In general, it is not possible to eliminate this risk because, in most circumstances under state law, leases must remain cancellable by the lessee. The insurance company seeks to mitigate non-appropriation risk to the greatest possible extent, primarily through the evaluation of four key areas: (1) the issuer's overall credit; (2) the facilities to be leased; (3) the final term of the debt; and (4) the imposition of legal requirements.

A further discussion concerning the risk of non-appropriation may be found in Chapter 7, under Annual Appropriation.

Facilities To Be Leased

The nature of the facilities to be leased is a key determinant in assessing the likelihood that the lessee will consistently budget, appropriate for and make the lease payments. It is therefore the key to the investor in evaluating the level of non-appropriation risk.

Of the factors to be considered in assessing the nature of the facilities, *essentiality* is the most important. It is assumed that appropriations are much more likely to be continued with respect to property or equipment that is considered to be most essential to the lessee's operations.

Essentiality is obviously a subjective term, and its meaning may change over time. The rating agency's or bond insurance company's definition of essentiality may differ significantly from that of the issuer, underwriter, or bond counsel. For the rating agency's or insurance company's purposes, an essential facility is defined as any facility that is considered to be so critical to the continued governmental operations of the lessee throughout the term of the certificates that the likelihood that the lessee will be compelled, by necessity, to make appropriations for lease payments is extremely high, regardless of the conditional nature of the lease obligation.

The rating agencies or municipal bond insurance companies have a very strong preference to rate or insure obligations issued to finance real property and improvements, rather than equipment. Their preference for real property-backed financings is based upon the fact that real property does not generally depreciate in value and, as collateral, can be expected to be marketable at a reasonable price in the event the lessee defaults. Further, the improvements on real property generally have a useful life that may extend well beyond the term of the debt obligation. Finally, the improvements on real property are less likely to experience the operational problems of equipment that might cause it not to be useable, and the risk and problems of obsolescence are generally minimal. Equipment may be financed under a master lease arrangement with cross default provisions for such real property as may be included under such a master lease.

As a rule, all of the component risks of non-appropriation increase as the term of the debt increases. Therefore, the bond insurance companies do not generally accept lease transactions with a final maturity longer than 20 years, primarily as a consequence of increasing political uncertainty and risk over time.

Statutory Authority for Leasing

The source of statutory authority for a governmental unit to enter into a lease purchase agreement varies in all States. The specific language may grant power to the unit to *contract, purchase, lease,*

lease or purchase, or may provide explicit authority to *lease purchase.* In some jurisdictions, there are comprehensive statutes authorizing lease purchase agreements that have numerous limitations and restrictions. At least one State authorizes the lease purchase of property by a municipality only from not-for-profit corporations or authorities. In a few states, no statutory authority to lease purchase exists, but case law has construed and up-held lease purchase agreements. In the State of New York, for example, the General Municipal Law permits municipalities to enter into lease purchase agreements for equipment, such as computers and vehicles, but such authority does not include real property. Therefore, in accordance with Dillon's Rule, the Comptroller of the State has opined that, under current law, real property lease purchase agreements are not within the powers of a New York State municipality.

Dillon's Rule, followed by many jurisdictions, states, in brief, that a municipal corporation possesses and can exercise the following powers and no others: first, those powers granted in express words; second, those powers necessarily or fairly implied in, or incident to, the powers expressly granted; third, those powers essential to the accomplishment of the declared objects and purposes of the municipal corporation, not simply convenient but indispensable. Any fair, reasonable, substantial doubt concerning the existence of a power is resolved by the courts against the municipal corporation and the power is denied. All acts beyond the scope of the powers granted are void. (Dillon, *Municipal Corporations* §55, 1st ed.)

In 1986, the American Bar Association published in *The Urban Lawyer*, Vol. 18, No. 1, a 50-state survey of legislation and case law concerning local government leasing. The survey focuses on counties, municipalities, and school districts, with respect to the authority of and consequences to such local governments of entering into lease purchase agreements. The survey also shows the impact of such a lease purchase agreement on any debt limitations imposed on the local government by state constitution, state statute, or local charter. In this respect, the research found a significant distinction in many states between a long-term lease purchase continuous agreement and a true lease or lease purchase subject to annual governmental appropriation and annual renewal.

Appendix 6 shows a summary of the results of this survey. While this summary constitutes a reasonable interpretation of the complex findings in each state, legislation and case law in many states do not lend themselves to an easy yes or no answer. Briefly, the following conclusion can be drawn from this survey:

	Yes		*No/None*	
Description	*Number*	*%*	*Number*	*%*
Ability to lease	50	100%	—	—
Public bidding for lease	38	76	12	24%
Interest rate limitations	17	34	33	66

Debt Limits	*Number*	*%*
Not debt, if annual appropriation	38	76%
Subject to debt limits	7	14
Leases not under debt limits	5	10
TOTAL STATES	50	100%

Legal Requirements of a Lease

There are basic structural requirements that should be included in the lease or associated legal documents. In summary form, they are as follows:

- *Perfected Security Interest:* A perfected security interest is required in all property and equipment to be leased. In the case of real property, a first mortgage lien or a leasehold mortgage that lasts for a period significantly longer than the final maturity of the debt is also required.
- *Triple Net Obligation:* The lessee's obligation to make lease payments must be triple-net in nature; i.e., absolute and unconditional, and not subjct to any offset, counterclaim or recoupment arising out of failure of any party in the lease transaction to perform its obligations. An opinion to that effect will be required from bond counsel.
- *Assignment to Trustee:* All right, title and interest that the lessor has in the lease payments and the leased facilities must be properly assigned to the trustee.
- *Reserve Fund Requirement:* In general, a maximum annual debt service reserve fund is required.
- *No Construction Risk:* The lessee's obligation to begin making lease payments must not be contingent upon completion of the facility. If a construction risk exists, an alternative is to provide

an acceptable letter of credit that would be used to redeem the debt if the facility is not completed by a specified date.

- *Cross-default Protection:* If the facilities to be leased include some that are considered to be less essential than others, there should be a cross-default provision whereby the lessee will be in default on all facilities if they fail to make lease payments for one or more of the leased facilities.
- *No Abatement:* The lessee's obligation to make lease payments must continue despite damage or destruction that makes either all or a portion of the facility unuseable; that is, abatement of lease rental payments is typically not acceptable.

In the event of a default on the payment of interest or principal by the issuer of bonds insured by a municipal bond insurance company, the insurance company makes the required payment directly to the trustee or paying agent bank. When such payment is made, the insurance company receives the interest coupons and/or the matured bond certificates, which become a lien on the issuing municipality in accordance with the insurance agreement. Payment on a future date then becomes a negotiated settlement between the insurance company and the issuer.

CREDIT ENHANCEMENT

In order to provide an adequate level of security for debt obligations, it may be necessary to pay for private credit enhancement, in addition to whatever the issuer may be willing or able to provide directly. The two predominant forms of outside credit enhancement available today are municipal bond insurance for tax exempt bonds and bank letters of credit for both tax exempt and taxable obligations. Bond insurance provides the highest ratings for the proposed financing but is very conservative in analysis. Letters of credit from a commercial bank are more expensive and provide a rating for the proposed bonds only as good as that of the letter of credit bank.

Municipal Bond Insurance

In the early 1970s, the insurance industry began providing irrevocable insurance to issuers of tax exempt municipal bonds. Because of the advantages to issuers in reducing the cost of borrowing in the marketplace, the municipal bond industry has grown to the extent of insuring over 25% of all new issues of tax exempt bonds in 1989.

Such insurance provides the bondholder a guarantee of the timely payment of principal and interest. The policy is granted to the municipal issuer only upon careful analysis by the insurance company and in exchange for a one-time premium. All of the companies writing such bond policies charge a premium based on a fixed percentage of the total principal and interest payments over the life of the bond issue, less such amounts of interest as may be capitalized from bond proceeds. Typical insurance premium costs range from 0.50 to 1.00% of total principal and interest over the life of the insured obligations.

Currently, there are four insurance companies providing bond insurance. The following shows each company and the ratings assigned to their claim paying ability:

Company	*Moodys*	*Standard & Poors*
Ambac Indemnity Corporation	Aaa	AAA
Capital Guaranty Insurance Co.	—	AAA
Financial Guaranty Insurance Co.	Aaa	AAA
Municipal Bond Investors Assurance Corp.	Aaa	AAA

Late in 1989 Municipal Bond Investors Assurance Corporation purchased Bond Investors Group, Inc., the parent of Bond Investors Guaranty, another bond insurance company that had been in business for several years. According to figures from Securities Data Co./Bond Buyer for the first nine months of 1989, the combined companies would enjoy a 52.6% share of the market in the municipal bond insurance field. Ambac Indemnity Corporation follows with an estimated 29.8% market share, Financial Guaranty Insurance Co. with about 15% of the market and Capital Guaranty Insurance Co. having the remainder.

Both Moodys Investors Service and Standard & Poors Corporation assign the ratings shown for a particular insurance company to the issuer's bonds, if an insurance policy has been written by that company.

All insurance companies have similar analytical policies and procedures they follow, to determine if a tax exempt bond is eligible for insurance. The type of analysis and the data examined will vary depending on the type of bond and the basic security.

The basic requirements for a general obligation bond, whether payable from limited or unlimited property taxes, include:

1. the clearly defined obligation of the issuer to levy and appropriate ad valorem taxes when and in the amount needed to pay debt service on the bonds;
2. the issuer having a diversified, well-balanced economic base, with satisfactory employment levels;
3. the issuer having a tax base evidencing continuing economic growth, as measured by population growth, new construction permits, or increasing valuations;
4. the proposed debt to be insured, when added to any outstanding obligations, appears manageable in relation to property values, income levels, and the total revenues available to the issuer;
5. the issuer demonstrating the practice of sound accounting principles in budgeting, financial reporting, and cash management;
6. the issuer demonstrating a consistent history of satisfactory tax collections, along with satisfactory enforcement procedures,
7. the fact that employee benefits and pensions are actuarially funded; and
8. that the bonds to be insured are payable serially or through mandatory sinking fund deposits in trust.

The criteria used by bond insurance companies to analyze special assessment districts or local improvement districts include the following:

1. the assessment liens created must be subordinate only to Federal tax liens, but must be co-equal to the lien for general property taxes;
2. the basis for assessments must be fair, objective and equitable;
3. provision must be made for an adequate debt service reserve;
4. unless the proceeds of the financing are used to acquire completed improvements at a fixed cost, the issuer must have sufficient funds or financing capability to cover significant cost overruns;
5. the issuer must covenant to initiate prompt foreclosure proceedings against delinquent properties;
6. if possible, the collecting agency should make payments directly to the bond trustee;
7. additional credit support may be required to secure any major assessment with a lien to value ratio less than 1:5, or composing more than 10% of the obligation; and

8. any legal challenges, claims, or adjudication with respect to the special assessment must be resolved.

Tax increment financing calls for a somewhat different set of requirements for approval, including:

1. certified tax increments for the most recent fiscal year must cover adequately maximum annual pro forma debt service;
2. such tax increment coverage must be determined on the basis of net tax increment, after allowance for any statutory or pass-through deductions;
3. an adequate debt service reserve must be provided;
4. the project area must show evidence of development potential, recent private investment and diversification; and
5. an appropriate additional bonds test must be included.

The requirements for revenue bonds may vary somewhat, depending on the nature of the enterprise project or system, but basically would include the following:

1. a rate covenant which provides an adequate debt service coverage from net revenues, with provisions to replenish the debt service reserve fund within five years or less;
2. an additional bonds test which provides adequate protection, with an earnings test preferably based on current financial performance;
3. a debt service reserve required to be maintained at a sufficient level;
4. historical trends of stable or increasing financial viability;
5. the economy of the service area is stable or improving; and
6. the bond maturity schedule is shorter than the estimated useful life of the facility or improvement being financed.

In the past, all bond insurance companies have shown a reluctance to write an insurance policy on the financing of a single, start-up parking project, unless the municipality provides an acceptable form of guarantee or pledge of general fund revenues. If the revenues from a large, well-established city parking system are pledged to a parking revenue bond, then an insurance policy may be considered, given the insurance company's satisfaction of the normal requirements of a revenue bond.

Until a few years ago, tax-exempt leases, which depended upon annual appropriations were not considered insurable. More recently, because of new entrants into the municipal bond insurance industry and more experience with annual appropriation risk, lease purchase transactions with strong security features have been eligible for insurance.

Letters of Credit

A letter of credit constitutes an irrevocable obligation of a bank, for a specific period of time, to make payments upon the demand of the entity to which the letter of credit is drawn, up to the total amount stipulated therein. It states the procedures to be followed to produce a draw down of funds and how the funds are paid by the bank to the holder of the letter of credit. It also makes provision for the renewal of the letter of credit, upon termination of the original term, solely at the option of the bank.

A letter of credit is always accompanied by and is made a part of a reimbursement agreement between the issuing bank and the entity to which the letter of credit is drawn. Basically, the reimbursement agreement sets forth the application of the entity or "borrower" for the letter of credit, the granting of same by the bank for a specific amount and term, the rates charged by the bank, and the obligation of the borrower to repay any amounts drawn down pursuant to the letter of credit. In the event of a draw down by the borrower, the reimbursement agreement classifies this amount as an "advance" and sets forth the obligation of the borrower to repay the advance, at the bank's lending rate over a specified period of time, from any moneys available to the borrower. In effect, any "advance" made to the borrower by the bank becomes a bank loan to the borrower.

In most instances, a bank issuing a letter of credit will demand a first mortgage on the property and equipment of the parking facility for which financing the letter of credit represents credit enhancement. Such a mortgage granted in favor of the letter of credit bank would take precedence over any mortgage granted to bondholders. Any letter of credit bank payments, made due to insufficient project revenues, would prevent a default in the payment of debt service on such bonds, thereby avoiding an event which might trigger a mortgage foreclosure by the bondholders.

Obviously, prior to issuing its letter of credit as enhancement to any parking bond issue a bank must be satisfied as to the basic credit quality of the financing and the parking facility, the revenues from which are pledged to secure the bonds. Similarly, when a bank issues a letter of credit to enhance a developer's mortgage loan or an IDB secured by a developer's lease agreement, it must be satisfied with the credit worthiness of the developer and the economic feasibility of the parking project being financed.

Commercial banks, both foreign and domestic, have provided letters of credit, standby letters of credit, standby bond purchase agreements, or other types of credit enhancement and liquidity facilities to municipal borrowers for decades. At present, there exists an estimated $80 billion of bonds issued by municipal borrowers, which rely on credit enhancement from foreign and domestic banks.

International Regulation

In an effort to standardize the treatment of letters of credit on banks' balance sheets and provide some regulation, banking agencies in 12 major industrial nations have proposed banking regulations based on an international bank regulatory agreement, known as the "Basle Capital Framework." The agreement (1) provides a universal definition of risk assets, which includes items not previously included in capital ratios, such as off balance sheet assets like letters of credit; (2) requires that minimum capital reserves for these risk assets be held by all banks; and (3) establishes a uniform reserve requirement for risk assets, which increases from 6 to 8% of weighted risk assets by the end of 1992. The countries who will comply with this framework and who are the major suppliers of letters of credit are Belgium, Canada, France, Germany, Italy, Japan, Luxembourg, the Netherlands, Sweden, Switzerland, the United Kingdom, and the United States.

The Basle Capital Framework's standard requirement that these credit enhancement vehicles have capital reserved against them and, more importantly, the quality of capital which must now be reserved against these credit enhancement vehicles will probably force banks to reconsider the pricing of letters of credit. In the past, letters of credit were priced by banks based on the credit risk of the transaction and as a means to develop banking relationships. All banks will now have to add to the price of letters of credit an appropriate return on capital and return on equity to maintain their profitability and their ratings. Banks typically target a 15 to 20% return on equity.

The direct effect of the Framework on corporate and municipal issuers will result from the Framework's accounting of letters of credit:

> There are presently no minimum required capital reserves for outstanding letters of credit in many countries, such as the United States. Under the Basle Capital Framework, each outstanding letter of credit will be counted as a risk asset (against which a bank must have sufficient capital reserves) in the following percentages of par amount outstanding: 20% for general obligation issues, 50% for municipal revenue bond issues, and 100% for industrial development bond and corporate issues.
>
> The minimum capital reserves required for letters of credit for general obligation, municipal revenue, and corporate and industrial development issues will be 1.45%, 3.625%, and 7.25%, respectively, of outstanding letters of credit by 1990, and 1.6%, 4%, and 8%, respectively, by the end of 1992.
>
> A minimum quality standard of capital which is reserved against letters of credit will be imposed. The minimum amount of core capital (which is mostly shareholder equity) that has to be reserved against letters of credit for general obligation, municipal revenue, and corporate and industrial development issues will be at least 0.725%, 1.625%, and 3.625%, respectively, of outstanding letters of credit by 1990, and 0.80%, 2%, and 4%, respectively, by the end of 1992.

Banks will need to reduce risk assets, such as letters of credit and raise capital reserves against outstanding letters of credit to comply with the Framework. As a result, corporate and municipal borrowers who use letters of credit may incur substantial expenses in the following three ways:

- Outstanding letters of credit usually have provisions under their reimbursement agreements, which require an automatic increase in letter of credit fees to compensate banks for capital requirements imposed on letters of credit.
- Once the Framework is fully effective, banks may increase their fees substantially on the renewal dates of outstanding letters of credit or choose not to renew the letter of credit at all.

- The broad definition of risk assets and the requirement of minimum capital reserves being established against these assets may put great pressure on the ratings of banks. If banks are downgraded, the interest rate on proposed debt enhanced by letters of credit issued by these banks will rise to compensate investors for the declining credit rating.

Objectives of the Framework

The Framework has been adopted by the participating nations with the stated purpose of improving the stability of the international banking system and of removing an important source of competitive inequality for banks which arise from differences in individual nations' supervisory requirements.

According to the Basle Supervisor's Committee report, the specific objectives are:

- A common framework for defining and comparing the capital positions of banking organizations from different countries, with emphasis on common stockholders' equity and appropriate recognition of other components of capital.
- A system for relating capital to banking risks, including items not previously included in capital ratios such as off-balance sheet exposures, in order to enhance the quality of capital as well as the level of capital.
- A schedule requiring minimum capital reserves for those risk assets held by all banks.
- Appropriate transitional arrangements designed to provide a reasonable amount of time for banks to bring their capital positions into conformity with the risk-based Framework.

Definitions of Capital and Reserves

The Framework defines the constituents of capital in two tiers: Core Capital (Tier 1) and Supplementary Capital (Tier 2). Core Capital or basic equity is the only element of capital common to all countries banking systems. Equity capital is readily accessible in statements published by banks and is currently the basis on which most market judgements of capital adequacy are made. Core Capital includes all issued and fully paid common stock, excluding preferred stock and published reserves from post-tax retained earnings. At least 50% of a bank's capital base must consist of Core Capital. Supplementary

Capital will include other forms of capital, such as loan loss reserves, preferred stock, subordinated debt, and revaluation reserves up to an amount equal to that of the Core Capital.

Although most banks have sufficient Supplementary Capital, the quality standard imposed by the strict definition of Core Capital will force most banks to raise common equity and/or shrink assets. Banks will need to evaluate all business activities which require capital reserve requirements. Investments by banks in corporate and municipal bonds and all letter of credit business supporting corporate and municipal issues will have capital requirements which will use up Core Capital availability. Because they will use up Core Capital, these activities will have to provide a return on equity commensurate with all other capital intensive banking businesses.

A weighted risk ratio in which capital is related to different categories of asset to off-balance sheet exposure and weighted according to broad categories of relative riskiness is used to assess the capital adequacy of banks. The Committee believes the risk ratio method will provide an equitable basis for international comparisons, allow off-balance sheet exposures to be more easily incorporated into the measure, and will not deter banks from holding liquid, low-risk assets. For example, a bank will not have to hold any capital (the 0% risk category) against such assets as cash or short-term government securities and it would have to hold the most capital (the 100% risk category) against such assets as commercial loans and letters of credit securing a tax exempt industrial development bond.

Municipal bonds are also affected by these risk weights in that any asset issued through a tax exempt entity, such as an investment in a municipal bond or a loan to a municipal issuer, will have to have capital reserved against it. Tax exempt bonds secured by a general obligation have a 20% reserve requirement, by a revenue pledge a 50% requirement and by a third party (such as an IDB) a 100% requirement.

There are presently no minimum required capital reserves for off-balance sheet exposures, such as standby letters of credit in many countries like the United States. The risk-based capital system brings all of these off-balance sheet items into the capital framework, with risk weights assigned to reflect the actual credit risk associated with each type of asset.

The new guidelines apply a credit conversion factor to the off balance sheet asset. Items that are direct credit substitutes, including letters of credit, have a conversion factor of 100%. After the conversion factor is applied to an off-balance sheet item, the result is risk-weighted. In this way, the off-balance sheet item is translated first into the credit-equivalent of an asset, then assigned a risk-weight, like any other asset.

The effect of this accounting for off-balance sheet items on corporate and municipal issuers will be that all letters of credit, standby letters of credit, bond purchase agreements, and participations in these forms of credit enhancement will be treated like a loan for capital purposes and have a conversion factor of 100%. The pricing of these services will have to meet the return on capital and return on equity standards of the bank, based on the amount of total capital and Core Capital which such credit enhancement obligations use up.

The effects of the Basle Framework will vary from country to country depending on the prior regulatory and banking practices followed in each country. The following discussion summarizes the effects of the new capital regulations on U.S. banks.

In 1985, U.S. bank regulators adopted uniform minimum primary capital-to-assets guidelines for all U.S. banks. These guidelines differ from the Basle Framework in two fundamental respects:

- The guidelines defined a much broader definition of capital, which included general loan loss reserves and intangible assets.
- The guidelines did not include off balance sheet items, like letters of credit, as assets which required capital reserves.

These differences have enabled U.S. money center banks to establish substantial reserves of at least 25% for debt of underdeveloped nations without violating the United States regulatory guidelines and encouraged them to move assets into off-balance sheet activities like letters of credits, etc. The FDIC recently reported that over one-third of all U.S. banks with more than $3 billion in assets had more standby letters of credit outstanding than they had capital.

As measured by the Basle Framework criteria, many United States money center banks are deficient in total capital and Core Capital. A study by the banking consulting firm, Keefe, Bruyette & Woods Inc. calculated the common equity-to-assets ratios of 197 U.S. banks. These ratios excluded goodwill, as specified in the Framework. Using June 30, 1987 data, 26 of the top U.S banks would fall short of the Core Capital requirement of 4% under the Basle Framework (see Table 5-1).

Table 5-1. Major United States Banks With Equity Ratios Under Proposed 4% Minimum Core Capital Requirements (as of June 30, 1987)

	NET TANGIBLE COMMON EQUITY RATIOS
BancOklahoma Corp.	(1.13)%
First City Bancorp Inc.	1.31
Manufacturers Hanover Corp.	1.56
Banks of Mid-America Inc.	1.74
Continental Illinois Corp.	1.77
BankAmerica Corp.	2.10
Citicorp	2.16
Chemical New York Corp.	2.17
MCorp	2.53
Wells Fargo & Co.	2.53
Chase Manhattan Corp.	2.54
First Republic Bank Corp.	2.72
Marine Midland Banks Inc.	2.92
Mellon Bank Corp.	3.01
First of America Bank Corp.	3.07
Lane Financial Inc.	3.15
Security Pacific Corp.	3.15
First Pennsylvania Banks Inc.	3.20
First Chicago Corp.	3.23
Irving Bank Corp.	3.37
River Forest Bancorp	3.52
Fidelcor Inc.	3.63
Equitable Bancorp.	3.79
Bank of New York Co.	3.90
First Interstate Bancorp	3.93
Northern Trust Corp.	3.98

Source: Keefe, Bruyette & Woods Inc.

These large banks need to issue new and/or reduce total assets to comply with the Basle Framework. Many regional banks and small local lenders are already at or above the 4% Core Capital and 8% total capital requirements.

A second effect of the Basle Framework on U.S. Banks will be that banking activities, such as providing letters of credit, which consume Core Capital and total capital under the Basle Framework, but did not under prior U.S. guidelines, and are booked as assets under the Basle Framework, but were not under prior U.S. guidelines, will have to be repriced. United States banks are obligated to their shareholders to maintain profitability standards based on return on equity and return on capital. Letter of credit fees will need to increase for all U.S. banks to maintain these profitability standards.

As of October, 1989, 4,535 U.S. tax exempt municipal bond issues are backed by letters of credit, according to Securities Data Co./The Bond Buyer. Of those, 2,676 with a dollar volume of $61.48 billion are backed by domestic bank letters of credit. The remaining 1,859 issues totaling $72.43 billion are backed by foreign banks. In the last three years, more than 40% of all the par amount of tax exempt securities backed by letters of credit have been issued by Japanese banks. These Japanese banks have undercut the domestic competition with significantly reduced fees, as well as marketing their high credit ratings to support the use of their letters of credit for credit enhancement.

Few U.S. banks have AAA ratings on letters of credit, as a result of a series of credit rating downgrades on domestic banks in the early 1980s, creating an opportunity for the well capitalized Japanese banks.

Table 5-2 shows the ranking of both foreign and domestic banks, with respect to the volume of tax exempt municipal bonds for which they provided letters of credit.

Table 5-2. Long-Term Public Municipal New Issues 1/1/90-12/31/90

RANK	BANK	$ VOLUME[1]	MARKET SHARE	ISSUES
1	National Westminster Bank PLC	887.5	7.6%	23
2	The Fuji Bank, Limited	810.1	7.0	21
3	The Industrial Bank of Japan Trust Co.	804.7	6.9	16
4	Sumitomo Bank	737.3	6.3	19
5	Union Bank of Switzerland Limited	686.6	5.9	17
6	The Sanwa Bank, Limited	668.6	5.7	17
7	Swiss Bank Corporation	488.7	4.2	24
8	Banque Paribas	347.5	3.0	8
9	Morgan Guaranty	317.2	2.7	8
10	Dai-Ichi Kangyo International Limited	286.1	2.5	16
11	The Yasuda Trust & Banking Co.	214.6	1.8	5
12	Mitsui Trust & Banking Co., Ltd.	204.3	1.8	4
13	Barclays Bank PLC	201.0	1.7	7
14	Mitsui Taiyo Kobe Bank, Ltd.	172.4	1.5	5
15	Mitsubishi Bank	161.9	1.5	5

[1]In millions
Source: Securities Data Co., Inc.

CHAPTER 6

TAXABLE FINANCING STRUCTURES

The options available for private parking operating companies or developers to finance parking facilities are more numerous than are available to local governments. Usually, developers are restricted in their financing options only by the economics of the project and the prevailing appetite of the investor market. In the event that a large credit-worthy parking company or development company elected to place its corporate credit behind a stand-alone parking project, the financing options would be increased. Several types of financing structures are discussed below; the use of one method over another depends on a thorough economic analysis, based on interest rates charged in the market and the overall fiscal policies of the borrower.

The ownership of private parking facilities has been characterized as another form of the real estate business. Excluding the nuances of operating any parking facility, which is a unique type of business unlike any other, the acquisition, financing, and ownership considerations of the private parking industry primarily track those of the real estate business. All of the financing structures discussed below are used for general real estate financing, as well as being available for parking "real estate."

CONVENTIONAL DEBT/EQUITY

The choice of financial structure for the developer of a new parking project is an integral part of the development decision-making process and is a decision that can have a significant effect on profitability. Developers may choose among a variety of financing alternatives that can be very different from both a cash flow and a legal perspective. It can be assumed that the developer will seek to obtain financing that is affordable and that maximizes the expected rate of return on equity.

The following types of financial structures for an income property, such as a parking project, are available in today's marketplace:

- Bullet loans;
- So-called thrift deals that feature negative amortization plus participations;
- Pension participation mortgages with developer or lender priorities; and
- Land sale/leasebacks.

Although there are an infinite number of variations on the various terms of each type of financing, the basic formats of these four types of structures are applicable to most income-property development situations.

The Bullet Loan

A "bullet (or partially amortized) loan" is the current version of a fixed-rate mortgage and is a commonly available alternative. Payments are based on a 25- or 30-year amortization schedule with a 10-year call provision at the option of the lender. Should the lender choose to call the loan, the mortgagee must satisfy the remaining loan balance, either by selling the property or by refinancing at the then current interest rate. The call provision reduces the lender's inflation and interest-rate risk because it reduces the effective mortgage term, while the 30-year amortization schedule does not reduce affordability relative to a standard fixed-rate mortgage. Indeed, competitive market pressures may force a lender to offer bullet loans at rates lower than those of comparable standard mortgages to the extent that the lender's interest-rate risk is reduced.

The Thrift Deal

One of the most common savings and loan deals involves a mortgage instrument that has an initial "pay" rate that is less than the note (contract) interest rate. Thus, for example, initial payments on a 10.25% mortgage may be based on a 9% pay rate. The difference between the 10.25% note rate and the 9% pay rate is added to the outstanding loan balance (accrued) and paid by the borrower when the property is sold or refinanced. Typically, the pay rate increases in annual increments during a 3- to 5-year period until it equals the note rate. This financing structure makes the loan more affordable for the developer without reducing the lender's expected yield. Of course, because of the negative amortization, the loan is riskier for the lender than a standard loan.

In addition to the interest accrual feature, the thrift loan is structured to give the lender an "equity-like" interest in the property by means of a participation in annual cash flow and/or a participation in sale proceeds. The participations are designed to help protect the lender from an unanticipated increase in inflation and interest rates and to allow the lender to share in benefits (and costs) that usually accrue only to the equity investor.

In recent years, increasing numbers of savings and loans associations have experienced the downside potential associated with equity anticipations. Some of the unsound loans that have caused the Federal Home Loan Bank Board to take over or merge a number of savings and loans have been equity participation loans. In the current environment of tighter regulation of lending practices, participation loans are being discouraged by S&Ls.

Pension Participation Mortgages

Many pension funds today actively finance commercial real estate investments. They have also become more active as equity investors in recent years. Although pension funds utilize a variety of different loan types, they favor a format in which they offer permanent below-market-rate mortgage financing (typically 75 to 150 basis points below market) in exchange for an equity type interest in the property (sometimes 50% of both operating cash flows and proceeds from sale). The reduced interest rate makes the loan more affordable, but the loan reduces the "upside" potential of the developer's return because the lender will share in any larger-than-expected increase in rents and resale values.

The Land Sale/Leaseback

The land sale/leaseback is a financing form frequently available from insurance companies. The lender purchases the land, which it subsequently leases back to the developer, and it provides the long-term mortgage financing for the improvements. The lender's cash receipts include both debt service on the mortgage note and lease payments on the land. Because the lender has a long-term economic interest in the land, it may gain from any unexpected property price appreciation. In addition, the lender frequently participates in the annual cash flows and/or in the proceeds from sale of the improvements. Under a true sale/leaseback, the developer (seller) relinquishes the economic interest in the land. However, depreciation deductions are retained on the improvements, which, along with the lease payments on the land and interest payments on the building mortgage, are generally fully deductible. Because of the benefits the lender receives from the land acquisition and cash flow participations, the mortgage rate is a below-market rate.

SELECTING THE BEST ALTERNATIVE

Unfortunately, the selection process is more complicated than merely choosing the financing alternative that is expected to produce the largest quantity of after-tax cash flows. Such quantities must be adjusted to reflect both their timing and relative risk. An investment decision rule that explicitly adjusts for the timing of the expected cash flows is net present value (NPV). But how does the developer compare risks?

The risk assessment process is simplified to some extent because operating cash flows do not vary with the method of debt financing. The method and amount of debt financing does, however, effect both before-tax and after-tax cash flows and the variability (riskiness) of these cash flows. For example, a higher loan-to-value ratio may increase the expected return on the project, but the increased leverage causes after-tax returns to be more sensitive to variations (from expectations) in operating cash flows and reversionary sales price. How much greater should the expected return be on the more heavily leveraged financing alternative to compensate for the increased riskiness of the developer's equity position? That depends directly on the developer's perception of how likely it is that the project will achieve projected income levels.

Debt service coverage ratios are not the binding constraints on maximum loan size. Although, in the past, coverage ratios frequently determined maximum loan amounts, recent declines in mortgage interest rates have caused the maximum amounts specified by traditional coverage ratios to be greater than the acceptable loan-to-value ratios available today in many markets. Consequently, because initial "affordability" is currently less of a problem for many developers, the relative popularity of participation mortgages is likely to decrease.

Comparing Present Value

When the after-tax cash flows produced by various financing alternatives are converted to present values with a 20% discount rate, the land sale/leaseback appears to dominate the other alternatives. Is the sale/leaseback deal the best alternative? Are the pension participations preferable to bullet loans? In making this decision, the developer must adjust for the fact that the pension participations may be relatively more risky than the bullet loan because of the larger loan amount (currently as high as 90% loan-to-value ratio) and because the smaller equity investment will cause actual returns to be more sensitive to variations in net operating income and resale prices. These required assessments of relative risk must be quantified in terms of the required equity discount rate. If 20% is the appropriate discount rate for, say, the bullet loan structure, then a minor adjustment to this rate would be required to reflect the varying degrees of risk inherent in the other financing structures.

Negotiating Points

Although, a land sale/leaseback may appear to be an attractive financing option, no financial structure is invariably the best. Available terms change on a weekly, and even daily, basis. Also, the lenders who prefer specific financial structures may not be interested in particular projects. For example, a thrift deal may be available for an office building, but not for a parking garage. Thus, the developer may have limited choices for a particular project type or location.

Because of the increasingly complex financing structures available, there are many different deal points to be negotiated; the most common found in the financing alternatives previously discussed, include the following:

- Loan amount
- Interest rate
- Origination fee
- Loan term
- Years of interest-only payments
- Amortization term
- Prepayment penalty
- Annual cash flow participation to lender
- Annual cash flow threshold priority to borrower (or lender)

- Reversion participation to lender
- Preferred return to lender (or borrower)
- Whether or not unpaid preferred returns are cumulative
- Terms of land sale/leaseback

Effective negotiation with potential lenders requires that developers understand the various deal points and how they affect both the lender's returns and developer's risks.

A developer who can estimate the lender's expected yield will understand which negotiations are most likely to be successful. For example, under some assumptions, a bullet loan may appear to be the least attractive alternative for both the developer and the lender. This suggests that it is unlikely that negotiations will produce a bullet loan agreement that is satisfactory to both parties. There are, of course, as many variations on these themes as creative negotiators can invent. Which deal is best for the parking developer can best be determined by a side-by-side analysis of the cash flows and by careful consideration of the different risks presented under each structure.

PARTICIPATING MORTGAGES

During the past decade, participating mortgages have become a widely used means of funding highly leveraged acquisitions of income-producing property, primarily through the growing investments of various pension funds. Many borrowers have been willing to share their profits with their lenders in return for high levels of financing at relatively low interest rates. Moreover, these "debt joint ventures" allow the borrower to retain most of the ownership benefits (like tax benefits, equity participation, and management control).

Traditionally, institutional lenders have been the source of participating mortgages. Recent trends, however, particularly the "deinstitutionalization" and "securitization" of real estate finance, have created a number of alternative lenders and have made possible new financing arrangements that involve some form of lender participation in the profits generated by the property.

In a participating mortgage, the lender's return includes both a contractually agreed-on amount of interest plus potential contingent interest expressed as a percentage of gross receipts or (more often) net cash flow, and/or a percentage of property appreciation upon sale or refinancing. The contractually agreed on "base" rate may be fixed or floating and is usually below the prevailing market rate. For example, a typical participating mortgage from a savings and loan association entitles the lender to a fixed amount of interest at a rate that may be as much as 2 percentage points below market, plus 50% of the "ups"—net cash flow after debt service and net proceeds from refinancing or sale.

Advantages to the Borrower

Participating mortgages benefit developers in the following four major ways:

Control. The borrower retains control over the property. Although there are often extensive negotiations about the circumstances in which the developer must consult with the lender, usually the lender must be consulted only with regard to major decisions. The borrower also retains title to and ownership of the property with the opportunity to repay the loan at some future point and then own 100% of the property.

Leverage. Although participating mortgages look increasingly similar to equity instruments, when carefully structured they are characterized as ordinary debt for accounting and Federal income tax purposes. Not only does the borrower achieve substantial financial leverage, but the debt also creates *basis* leverage, enabling the borrower to buy more depreciable basis for each dollar of equity invested.

Lower debt service costs. Because the base interest rate is usually below market, the level of monthly payments is also lower. The lower debt service costs permit higher levels of leverage. Less of the developer's own funds must remain tied up in the property. Lower debt service costs may also reduce the property's economic break-even point.

Tax benefits. The borrower retains ownership of the property and the depreciable tax basis is increased by the amount of the loan. All the depreciation deductions inure to the borrower/owner. In addition, interest payments, both fixed and contingent, are deductible. If the at-risk rules remain intact and the existing tax brackets prevail, higher write-off ratios may be necessary to attract outside equity. Flexible participating loan terms and lower debt service costs could allow borrowers the latitude to structure such deals.

PARTICIPATING LOANS EASE ACQUISITIONS

The participating loan minimizes the investor's cash investment in a number of ways. It is not uncommon for a lender who receives a substantial equity participation to offer 100% financing (or close to it) with a relatively low base interest rate. The lender is especially likely to do so, if a reputable appraisal reveals an undervalued property. Thus, if an enterprising developer locates an undervalued property, or one for which the value could be enhanced at a relatively low cost, the opportunity exists for an equity-free investment. For example, the developer could obtain an option to purchase the property from the seller, or, alternatively, to execute a purchase agreement that contains a liquidated damages clause (with the stipulated damages roughly equivalent to the price of an option). Either arrangement gives the developer time to approach a lender (ideally, with a favorable appraisal in hand) and negotiate a participating mortgage. (If time runs short, the developer can acquire the property with a bridge loan, and then refinance with a long-term participating loan.) If the lender accepts the favorable appraisal, the loan-to-(appraised) value ratio often warrants near 100% debt financing of the purchase price. Even where properties are not undervalued, lenders will often accept the less secure financial position of a high-rate loan, if the expected equity kicker sufficiently "sweetens" the deal.

New Sources of Participating Loans

Recent developments have opened up new sources of real estate financing through the evolution of entities that attract and channel capital into real estate. The new developments are commonly called "deinstitutionalization" and "securitization." With the general deregulation of the nation's financial institution structure, intermediaries and advisors of various types now compete with institutional lenders as sources of funds and makers of financing decisions for real estate developments. In addition, large developers and noninstitutional investors issue bonds, notes (even Eurobonds and Euronotes), commercial paper, partnership interests, and Real Estate Investment Trusts (REIT) interests, all backed by commercial mortgages to raise funds from the investing public and noninstitutional "lenders." Although many of these new financing arrangements involve fairly low-risk properties and straightforward debt structurings, some notable exceptions have used participating loan arrangements and/or have entailed high degrees of leverage.

Mortgage Loan Partnerships

In the last two years, publicly offered mortgage loan limited partnerships have emerged as a welcome alternative for sources of permanent financing. Rather than making equity investments in real estate, these limited partnerships lend secured money. Currently, there are more than thirty such partnerships either actively lending or in the SEC registration process. Sponsored by such major concerns as JMB, Balcor, VMS, Consolidated Capital Corp., and Paine Webber Co., these entities plan to raise between $1.7 billion and $2.4 billion for mortgage investment.

Mortgage partnerships are able to price their loans more aggressively than corporate entities because loan profits flow through to the unit holders tax free. In addition, most limited partner loan programs offer some form of lender participation in the profits. Thus, mortgage partnership loans offer the following features to developers: (1) lower base interest rates; (2) significantly lower front-end fees; and (3) lender equity participation of 25% or less. Again, the lower debt service costs may permit higher levels of leverage and facilitate minimum cash investment by the borrower. The borrower also retains a larger share of equity participation and the other ownership benefits previously discussed (control, leverage, etc.).

Mortgage-Backed Bonds

In recent years, there has been much discussion of the securitization of commercial mortgages. In 1984, Standard & Poor's developed a rating system for bond issues backed solely by commercial real estate. (Rated issues appeal to a much larger market than nonrated bonds because many investors, particularly Europeans, will not buy an unrated security.) Optimists predict that billions of additional dollars for long-term financing, at interest rates that are three-eighths to one percent below typical private mortgage financing, will be funneled into real estate by rated (and some unrated) issues of commercial mortgage-backed securities. Instead of borrowing from traditional sources, the real estate investor will be able to go directly to the public capital market.

Three typical mortgage-backed debt issues which have been sold, are: (1) American Express, for its World Financial Center corporate headquarters, (2) Olympia & York (59 Maiden Lane), and (3) Fisher Bros. (605 Third Avenue), all located in New York. These are obviously conservative deals involving single-property loans backed by the security of major firms. Conservatism will be the keynote as long as Standard & Poor's requires at least a 75% loan-to-value ratio for an "A" or "AA" rating. It is therefore unlikely that a rated issue will ever be used in a highly leveraged acquisition or refinancing.

Some of the unrated, privately placed bond issues, however, have been highly leveraged, although these issues are usually more costly to the borrower than their rated counterparts. Equity participation arrangements are negotiable in the private placement market and may serve to defray the higher costs.

Mortgage Real Estate Investment Trusts (REITs)

Mortgage REITs have also reemerged in recent years as an important source of real estate financing. These trusts use the funds they raise from investors to make mortgage loans to real estate developers, and a substantial portion of these loans are participating mortgages. Consolidated Capital Corporation has sold over $1 billion of REIT shares, the proceeds of which are going into high-yield participating mortgages. ICM Property Investors and Mellon Participating Mortgage Trust are also large participating mortgage REITs. Many REITs, still mindful of the debacle of the late 1970s, insist on conservative investments with comparatively low leverage. Others, however, have participated in riskier investments when equity participation agreements have improved their position. As competition becomes keener in the future, higher-leveraged deals may become acceptable again.

Again, a borrower might also consider forming their own mortgage REIT to raise needed funds. This approach was taken in a refinancing of New York's Rockefeller Center. A REIT was formed and shares sold to execute the refinancing. The proceeds of the offering were invested in a long-term loan secured by Rockefeller Center, eventually convertible to a participating interest.

A developer seeking to acquire income property with a minimum cash investment while retaining a high degree of ownership benefits should consider participating mortgages. In addition to traditional lenders, today's market offers several new sources of participating financing to borrowers willing to share their profits with lenders in order to obtain higher levels of leverage and lower debt service costs. Of course, not every deal warrants large offerings of participating mortgage bonds, and not every borrower need ponder the possibility of forming a new mortgage partnership. For smaller acquisitions, institutional lenders may still be the least costly, and often the only, source of participating loans.

Enterprise Capital Fund

In 1987, following the demise of most tax exempt industrial revenue bonds as a result of the provisions of the 1986 Tax Law, the National Association of State Development Agencies established American Development Finance Inc. (ADF), a not-for-profit corporation, located in Washington, D.C. ADF has created a capital pool program called the Enterprise Capital Fund (ECF), offering low-cost financing to small and medium sized businesses. ADF has raised funds for this purpose by selling commercial paper into the marketplace, secured by a bank letter of credit. According to ADF policies, the capital will be used mainly to make loans to finance small business growth and other projects that would contribute to job creation and economic development, when tax exempt industrial revenue bonds are unavailable or inappropriate.

As each loan from ECF to a borrower is approved, ADF will sell an equal amount of commercial paper through an underwriting by investment banks. Each issue of commercial paper is secured and guaranteed by an irrevocable letter of credit issued by the Paris-based Banque Indosuez. Based on the bank's credit, the commercial paper will carry the highest short term ratings (A−1+/P−1), thus achieving the lowest short term interest rates, as those rates vary from time to time. The term of the commercial paper issued by ADF varies from 15 to 90 days, depending on market conditions, with an average of about 45 days at any one time. As is typical with commercial paper, each issuance is rolled-over or remarketed at maturity, until funds become available from the loan borrower to repay the principal of the loan on a predetermined schedule.

Each loan made by the ECF must be sponsored by a development agency on the state, county, or local level. Such a development agency might be an industrial development agency, economic development commission, or a local economic development not-for-profit corporation sponsored by a local government. The loan can be instituted in one of two ways:

1. The development agency issues a taxable industrial development bond, which is purchased by the Enterprise Capital Fund;

the development agency then loans the proceeds to the borrower, as it would for a tax exempt IDB.
2. The development agency signs a statement of support (non-financial) for the ECFs financing of the project in question; ADF then enters into a loan agreement and lends the proceeds of the EDF loan directly to the borrower.

The choice between these two mechanisms may be made by the participating development agency. In neither instance is the development agency's determination of public purpose subject to criteria dictated by Federal legislation. The ECF loan program can be used by a development agency as a stand-alone project program or as a complement to other programs underway or to be developed.

At least 55% of each project loan must be backed by an equal value real property mortgage representing the land or building of the project being financed. Beyond this real property requirement, there are few limitations with respect to the types of projects, including parking garages, which can be funded through the ECF. Subject to final approval on individual projects by the ADF Board of Directors, new construction, renovation, acquisition of land, equipment and facilities, refinancing, engineering, and interim or permanent financing are all possible. According to the ADF, minimum project size is about $700,000, with projects under about $3 million subject to certain minimum up-front costs. There is no maximum project size. In addition, the ECF loan may constitute 100% of a project's cost, in effect, not requiring any equity to be put up by the borrower. Except in rare instances, each loan made by the ECF must be backed by credit enhancement in one of the following forms:

1. a bank letter of credit,
2. a local government guarantee or lease, or
3. a private insurance policy.

Local banks which provide the borrower with a letter of credit must be strong, generally $500 million or more in assets, but they need not be rated. Similarly, the local government guaranteeing the loan or leasing the project, should be credit-worthy. Whether rated or not, all credit enhancers must be accepted by Banque Indosuez.

The rare instance where credit enhancement might not be necessary, would be for a borrower which is a strong business enterprise, with at least 5 years of profitable operating experience, a good balance sheet, and an investment grade long-term debt rating.

Where a local bank provides a letter of credit to guarantee the borrowers loan, it will make the determination of the principal amortization schedule permitted, subject to the approval of Banque Indosuez, and will require a first mortgage on the project being financed with the ECF loan. The local bank will require the execution of a reimbursement agreement between the borrower and the bank, to be activated in the event its letter of credit is drawn upon to prevent a default on the ECF loan. In such event, the local bank's first mortgage takes precedence over the mortgage granted in favor of ADF.

Where a local government guarantees the borrower's ECF loan or leases the completed facility from the borrower, with a lease rental equal to debt service on the borrower's ECF loan, the guarantee or lease rental must be assigned to ADF as security for the borrower's loan. Local government guarantees are usually general obligations of the municipality extending over the life of the loan by the borrower. In such instances, the basic credit of the local government is utilized as the credit enhancement vehicle. If a lease of the project to the municipality constitutes the credit enhancement, the following additional criteria must be determined by ADF:

1. Is the lease a continuous obligation of the local government or subject to annual appropriation and possible termination?
2. Is the source of payment of the annual lease rentals restricted to project net revenues, payable from other enterprise revenues, payable from limited excise taxes, or a general obligation pledge of payment?

Obviously, the stronger the credit security pledged by the local government behind a lease agreement, the greater chance exists of approval by ADF.

If the borrower can purchase and assign to ADF as credit enhancement, a policy of insurance in an amount sufficient to cover the principal of the loan and interest for a period of perhaps 90 days, ADF would readily accept the borrower's loan application, provided the insurance company was well rated.

Whether enhanced by a letter of credit, a policy of insurance, or a governmental lease or guarantee, the amortization schedule of the borrower's loan from ECF may be as long as 20 years, depending on the policies of the credit enhancer. ECF makes its loans to borrowers based on prevailing taxable commercial paper rates. The all-in effective rate to the typical borrower, including the up-front ADF charges of about 2.70% of the face amount of the loan, would be significantly below the prime rate. In addition, the borrower would have to pay the annual cost of credit enhancement, if any, such as the local bank letter of credit.

The major disadvantage of the ECF borrowing program is the variable rate risk borne by the borrower. Each monthly payment made by the borrower to ADF, consisting of interest and amortization of the loan, will be adjusted to reflect actual current commercial paper rates. While commercial paper rates have consistently averaged a level significantly below prime rates, there have been instances where a reverse yield curve is present in the market. At such times, it is possible that short term rates, such as commercial paper, are higher than long term, fixed rates. Usually, such an anomaly is short lived, but the ECF borrower could be subjected to an indeterminate period of relatively high monthly interest rates.

The rate charged by ADF is composed of the current commercial paper rate, plus an annual charge of approximately 0.75% to cover the cost of the Banque Indosuez letter of credit for the ADF commercial paper, trustee's fees, and administrative costs. On average, however, over the life of a 20-year loan, a borrower could expect to pay a total rate of interest to ADF of anywhere from 0.50% to 1.00% below prevailing prime rates of major commercial banks. This rate, however, does not include the borrower's cost of providing local credit enhancement, if any.

For ECF borrowers concerned about extreme fluctuations in commercial paper rates or those wishing to fix their monthly loan payments, ADF does offer an option which provides for a "cap" on the borrowing rate. Through the investment banking firm which markets its commercial paper, ADF has devised a Protected Rate Program alternative, which works as follows:

- The borrower enters into a loan agreement with ECF which stipulates an amount, term, and amortization schedule. The borrower's monthly debt service obligations are set at a fixed rate (the "Fixed Rate") comparable to an alternative fixed rate financing for that borrower given the specified loan term. At the closing on the loan, the investment banking firm, on behalf of the borrower, will purchase an interest rate cap (the "Cap") in the amount of the loan, and at a strike price equal to the Fixed Rate less annual fees payable. The term of the Cap is maximized at three years.

- Upon receipt of each monthly payment, the difference between the Fixed Rate payment and the underlying commercial paper obligation of ADF plus all periodic fees payable, will be placed in a variable rate subsidy account (the "VRS Fund") by the Trustee. The funds that accumulate in this VRS Fund during the term of Cap, together with investment earnings thereon, should be sufficient to provide the borrower with a high degree of interest rate protection during the period from the end of the term of the Cap until the maturity of the loan.
- Specifically, upon expiration of the term of the Cap, the borrower will still make monthly payments at the Fixed rate. In the event that the underlying commercial paper rate plus fees payable rise to a level higher than this Fixed Rate, the VRS Fund would be accessed to make up the difference. The effect of the VRS Fund, then, is to provide a cushion to the borrower in the event of a rise of commercial paper rates. To the extent that the actual average commercial paper rate is such that the VRS Fund is never depleted, the VRS Fund would be held until the point that its balance was just equal to the principal outstanding on the loan. At that time, the remaining principal would be redeemed with the VRS Fund balance, thereby passing on to the borrower the benefit of the lower commercial paper rate over the life of the loan.
- The Borrower must assume the risk of total depletion of the VRS Fund before the full term of the loan program is completed, due to a possible persistence of unexpectedly high short term rates. To the extent that the Fixed Rate was insufficient to cover debt service, the borrower's loan rate would be increased to pay the actual underlying commercial rate plus all fees payable.
- At such time as the commercial paper rate plus fees went below the Fixed Rate, the borrower would continue to pay the Fixed Rate and the VRS Fund would begin to be replenished. However, in comparing this interest rate risk to the historical averages of the commercial paper market, the risk is not high, especially when compared with the benefits the program offers.

With respect the financing of parking facilities, the ECF loan program offers 100% debt financing at rates somewhat higher than tax exempt rates, but with none of the restrictions imposed by tax exempt bonds. Because the ECF loan is provided from taxable commercial paper, all restrictions imposed by the 1986 Tax Law on tax exempt bonds, concerning commercial operators, long term space leases or corporate financial guarantees are not applicable.

Taxable Certificates of Participation

Taxable certificates of participation (COPs) may be issued and sold into the marketplace to finance parking projects, if one of the following prerequisites are present:

1. A parking facility is built by a developer for a municipality but does not meet the public purpose criteria because:
   - more than 10% of the spaces will be utilized for private or corporate restricted parking;
   - more than 5% of the proceeds of the financing will be used to
   - the parking facility is constructed adjacent to and fundamentally for a corporate-owned office tower, hotel, or other corporate traffic generator which constitutes the major source of parkers.
2. A parking facility is built by a developer for lease or lease-purchase of a corporate entity, such as an insurance company, hotel, private housing development company, commercial shopping center, or industrial company.

Taxable COPs may be secured by the payments from either a true lease or a lease purchase agreement; in either instance, however, the following elements of security must be present for an acceptable level of marketability:

1. a lease term not less than the term of the COPs;
2. an acceptable bank letter of credit guaranteeing the payment of interest and principal of the COPs, with a draw down to pay off the COPs in the event of default, termination or non-replacement of the letter of credit upon its expiration;
3. if no letter of credit is present, then the lease must have a separate guarantee of a strong corporate credit (the developer or the lessee) which can be invoked in the event of default or termination of the lease for any reason;
4. a little-used method of providing acceptable credit enhance-

ment to an issue of COPs is by depositing with the COPs trustee collateral in the form of U.S. Treasury bonds, high grade municipal bonds, or well-rated bank CDs, to the extent of 120 to 140% of the principal amount of the COPs, by either the developer or the lessee; interest on the collateral is retained by the owner or, if put up by the lessee, can be used to offset lease rental payments; if not used to cover a default, the collateral is returned to the owner upon the retirement of the COPs.

In a true lease, title to the parking facility stays in the name of the developer/builder. The lease to the lessee can be a triple net lease, where the lessee pays all of the expenses of operation and maintenance and the lease rental payment covers the COPs interest and principal, as well as property taxes and trustee fees, or the lease may be something less, where the developer pays the cost of extraordinary maintenance and repairs, billing the lessee for such expenses as paid. In a true lease, the lessee has the option to purchase at any time at the fair market value of the parking facility, as determined by an independent appraisal, even after the retirement of the COPs indebtedness. In addition, the developer/owner can take the permitted depreciation on the facility.

In a lease purchase, the trustee holds the title to the parking facility for and on behalf of the lessee, subject to the lien of the lease purchase agreement. Usually the lease is a triple-net lease, where the lessee assumes all expenses, including operation and maintenance, property taxes, etc., plus paying the lease rental equal to the principal and interest on the COPs, plus trustee fees. The lessee has the right to purchase the parking facility, during the term of the COPs, by redeeming for cash the balance of the COPs outstanding, whereupon the title held by the trustee is transferred to the lessee lien-free. Such would be the case also when the COPs were retired according to schedule. In either instance, the operation of the parking facility can be accomplished by the lessee directly or through a management agreement with a parking operating company.

INTERIM FINANCING

Interim financing for privately developed parking projects usually is available from two sources, (1) internally generated corporate funds or, (2) a bank line of credit. In either instance, the developer would expect the reimbursement of corporate funds or the retirement of a bank loan, with interest, from the proceeds of the permanent financing expected to be done on completion of the project.

The use of existing corporate funds for preliminary project expenses may seem to be the least costly form of interim financing, but may not be so. If the total amount anticipated to be spent over the development period is such a level that would strain the resources of the company, to the extent of jeopardizing normal and prudent operations, then such a course could end up to be very costly in other areas. Also, if such corporate funds are available to the extent required, it may be that the return on the internal investment of such funds could exceed the cost of borrowing such funds from an outside source, i.e., resulting in a negative arbitrage situation. It is not unusual for the preliminary planning and feasibility work to be paid from internal funds, with the entry into the design, acquisition and construction phase being the trigger to go outside the company for interim financing, That point might also be the trigger for a "go-no-go" decision to be made for the continuation of the project.

External interim financing for privately owned parking projects can be arranged usually in an amount predicated to cover all preliminary and construction costs to completion. Depending on the size of the developer, start-up operating costs or working capital for the project may also be included as part of the aggregated interim financing requirements. A bank supplying the line of credit constituting the interim financing will require at least the following from the developer/owner:

- a general corporate guarantee or pledge of payment,
- a first mortgage on the land and the improvement thereon,
- a pro forma statement of the projected earnings and expenses of the project, and
- a projected schedule of draw down on the line of credit during construction.

The corporate guarantee or pledge of payment may or may not be of significant value to the bank analyzing the proposed line of credit. Frequently, developers will attempt to insulate their basic operating company by creating another corporate entity to be the owner and developer of a specific parking project. Having no stream of income and no assets, other than the projected parking project as it is constructed, a bank might be reluctant to grant such a line of credit.

In such an instance, the developer will attempt to obtain a "take-out" commitment from a permanent lender, to assign to the line of credit bank as additional security. If a credit worthy corporate guarantee is made available to the bank, such a take-out commitment would not necessarily be required; it would add to the security provided by the borrower and probably result in a lower cost of borrowing.

The requirement for a first mortgage on the land and improvements is a necessity. In the event the developer defaults on the line of credit obligation or files for bankruptcy, the bank can promptly foreclose on the property and take ownership. If such an event occurs during the process of construction, the bank has the option of completing the project, commencing operations and marketing for sale a functioning project or selling the incomplete project to another developer for the best price obtainable.

The pro forma statement of income and expenses of the project and the projected draw down schedule serve to assist the bank to analyze the economics of the project, independent of the credit worthiness of the developer. Other factors which would be analyzed by the bank include the overall construction price, as to its reasonableness in the local area, the reputation of the developer and the construction contractor, the attractiveness of the site for a parking project and the experience of the proposed operator of the project.

The discussion concerning Estimated Construction Disbursements in Chapter 11 includes the typical parking garage construction draw down schedule. Based on the data shown, a developer can calculate with a reasonable degree of accuracy the draw down schedule presented to the line of credit bank and, using the negotiated borrowing rate demanded by the bank, can calculate the total cost of the interim financing. In the example used in Chapter 11, the aggregate cost of interim financing for a $25 million hard cost construction project amounted to 10.6% of those hard costs.

CHAPTER 7

TAX EXEMPT FINANCING STRUCTURES

Bond financing by local governments for public parking facilities is governed by state statutes in all states and frequently by charters and/or local laws. As an example, some states do not permit local municipalities to issue revenue bonds for parking facilities. Other states are more flexible and permit revenue bonds, special district bonds, tax increment bonds, and a variety of lease purchase obligations. Some of the financing structures which can be used to fund the capital cost of new parking facilities by a municipality, are described below.

GENERAL OBLIGATION BONDS

Statutes in all of the states govern the issuance of general obligation bonds by municipalities, for financing parking facilities or any other capital project. Many such statutes govern the specific terms of general obligation bond issues, with respect to the maximum term of bonds, the annual principal requirements, redemption provisions, and the method of sale of such bonds. In addition, most states permit municipalities to issue bond anticipation notes, in anticipation of permanent funding, in order to ascertain the final costs of construction of a capital project, prior to the issuance of long term bonds. Many states require that general obligation bonds must be sold at public competitive bidding and many require a majority vote of the electorate at referendum before an authorizing bond resolution may be enacted by the governing body. As an example, the states of New York and New Jersey require a specific formula for the amount of principal of a bond issue, which can be retired annually over the life of the bond issue.

General obligation bonds will obtain the lowest possible interest rate or cost of borrowing for any given municipality. Because the full faith and credit of the municipality is pledged to such bonds, the rate of interest will reflect the best that the community has to offer. The only way for a municipality to improve on its own full faith and credit pledge to a bond issue, is to purchase municipal bond insurance. Such insurance will achieve a AAA/Aaa rating for that issue of bonds, which may be a higher rating than the pure general obligation bonds of the issuer, without insurance.

Care must be taken, however, in the issuance of general obligation bonds to finance parking facilities. Not only must all of the public purpose provisions of the 1986 Tax Law be observed, to preserve the tax-exemption of the bond issue, but the implications of such bonds being added to the outstanding statutory debt of the municipality must be observed in comparison to the fiscal policies of the municipality with respect to other, non-revenue producing capital improvement bonding requirements which the community may be facing.

SPECIAL ASSESSMENT BONDS

In many states, municipalities have the ability under state statutes to issue special assessment obligations; in some states such obligations also bear the full faith and credit of the issuing municipality. A special assessment bond is payable primarily from a special assessment within a specific benefitted district, over and above the regular property tax levy. Those special assessment districts created by the governing body for the financing of parking facilities usually include a downtown area whose property owners will benefit from the presence of the parking facility being financed. The theory behind such special assessment financing is that a parking garage in a downtown location will benefit merchants and other property owners in its area of influence and that those property owners would be willing to subsidize the operations of that garage to keep user parking fees to a minimum. While this form of financing is more prevalent in the far western states, it is not used to any great extent in the east or middle west. The credit implications of such special assessment financing are highly dependent upon the size and area of the special assessment district and the ability of such a district to levy an unlimited tax assessment on property owners. If, as is the case in some states, the general obligation of the municipality is pledged, in addition to the special assessment, then the above-mentioned credit considerations become moot. Frequently, however, opposition by the benefitted property owners to the creation of such a special assessment district, relegates this form of financing to oblivion.

TAX INCREMENT BONDS

Tax increment financing has been used primarily in Arizona, California, Illinois, and Florida, but, where authorized pursuant to implementing state statutes, may be utilized in many other states. Tax increment bonds are payable from a highly segregated form of ad valorem property taxes, usually on properties in a specified area within a municipality. It is used in conjunction with major urban renewal or development projects in a physical area usually not less than a square city block and frequently much larger. By ordinance, the municipality establishes a base year property assessment within the area. Any increase in these property assessments over the base year assessment, as a result of the urban development as constructed, becomes the assessment base for the segregation, at the regular tax millage, for tax revenues pledged for the payment of tax increment bonds. In some states, the legislation authorizing tax increment bonds requires that the full faith of the municipality also is pledged to such bonds, in the event the tax increment receipts are insufficient to meet bond debt service. Such tax increment bonds secured by the municipality's full faith and credit should have the same credit rating as the general obligation bonds of the community. However, tax increment bonds not enhanced by the full faith of the municipality may have a distinctly different credit standing. The following problems could arise:

- If tax increment bonds are issued too early in a new project's development schedule, the investor bears a risk of the developer's failure to complete, as projected, the scheduled development.
- If the area included for the levy of the incremental property tax is limited in size to one project of one developer, the security of the tax increment bonds in effect becomes an obligation of the developer, good or bad, and may jeopardize the tax exemption of such bonds under the 1986 Tax Law (security interest test).

In effect, a new parking facility proposed to be financed by a municipality with tax increment bonds should be of a size and location that will serve a larger area than one development project, and the focal point development project should be well under construction, with financing commitments sufficient for project completion.

PARKING REVENUE BONDS

Revenue bond financing for public purposes began in the U.S. in 1897, when the city of Spokane, Washington, issued $350,000 of waterworks revenue bonds. The factors and rationale which caused Spokane to pioneer in this new field of public financing are the same that have caused the subsequent expansion of revenue bond financing throughout the country, i.e., the evasion of legal restrictions governing the issuance of general obligation tax bonds or the imposition of ad valorem property taxes to pay for such bonds. During the 1930s, the Great Depression was instrumental in a significant expansion in the use of revenue bond financing. Cities and states with severely depressed tax budgets turned to self-liquidating, user charge methods of financing capital expenditures for public works. Currently, revenue bond financing of all types constitute about 70% of the total annual volume of new tax exempt bond issues.

Basic Security for Revenue Bonds

When a municipality issues parking revenue bonds, the primary legal security to the ultimate bond holder is the authorizing bond resolution or, in some instances, the trust indenture between the issuing municipality and a trustee bank representing the bond holders. In either case, that bond contract must have certain minimum provisions, (1) to protect the interests of the bond holder with respect to the disposition and holding of funds and the proper collection and flow of revenues, and (2) to give adequate authority to the municipality to operate and maintain the parking facility and charge sufficient rates to meet its obligations.

The basic provisions which must be included in any parking revenue bond resolution or indenture, are described in detail in Chapter 5.

Project Financing

Parking revenue bonds secured solely by the revenues from a single, stand-alone, municipally-owned parking facility are not acceptable in the market today, at any reasonable tax-exempt rate, unless clear, irrefutable evidence is presented to indicate the existence of an adjacent traffic generator of a magnitude that will produce a debt service coverage from net revenues of close to 2.00 times, at local market rates for parking. Lacking such a traffic generator, such revenue bonds will need some other form of host municipality subsidy or guarantee, in order to achieve an investment grade rating.

Parking System Financing

Successful parking revenue bonds (payable solely from parking revenues) outstanding today, are secured by a variety of parking facilities within the municipality, including off-street garages and lots and on-street meters. Given a reasonable history of efficient operation, good enforcement practices, adequate market rates for parking and reasonably protective covenants in the bond resolution or indenture, pure system parking revenue bonds can be marketed at acceptable rates. The revenue history should show a coverage of annual debt service, including the proposed new financing, from net revenues, of at least 1.30 times in each of the preceding three years. The revenue projections, including the new facility to be financed, should show a similar annual coverage of at least 1.50 times over the projected life of the new bonds.

In most instances, where system parking revenue bonds can attain an investment grade rating on their merits, a policy of municipal bond insurance will prove to be economically advantageous by reducing the annual borrowing rate.

LEASE REVENUE BONDS

Lease Revenue Bonds are in many ways structured the same as certificates of participation, except that the lessee is another tax-exempt agency with the statutory powers to construct, finance, and lease parking facilities to another municipal entity. Frequently, in an effort to exclude parking bonds from the debt limit of a municipality, the governing body will turn to another municipal agency, such as a parking authority, industrial or economic development agency, urban renewal agency, or a like pseudo-municipal agency controlled by the governing body, to do the financing. Such an agency usually will have more flexibility in bond issuance and in construction contracts under state statutes, than the host municipality. By such an agency issuing the required debt to finance a new parking facility and leasing the completed facility to the host municipality, the obligation of the municipality to meet lease rental payments, i.e., debt service on the debt issued by the agency, becomes a contingent liability on municipality's balance sheet, dependent upon the nature of the lease agreement and the security of the lease rental payments.

Again, similar to COPs financing, the credit standing of such lease revenue bonds depends on whether the municipality's lease rental payments are continuous or subject to annual appropriations. As a continuous lease payment over the life of the lease agreement, payable from general funds, the lease obligation becomes a form of debt to the municipality. However, if it is subject to annual appropriation by the governing body of the municipality, then it is only a liability during that year of appropriation. As discussed under COPs financing, the credit implications of a continuous pledge of lease rental payments vs. annual appropriations are obvious. See Appendix 9 for a chart showing the structure of a lease revenue financing.

ECONOMIC DEVELOPMENT AUTHORITY (EDA) BONDS

A method of tax exempt financing of parking facilities is available through the use of a local or state economic development authority (EDA) or industrial development agency (IDA) provided:

- such authority or agency has the power to issue long term indebtedness for parking facilities, and
- such authority or agency has the power to lease such facilities to its host municipality.

Given such powers, an economic development authority could take the place of a developer/lessor, and would issue lease rental revenue bonds in its own name, instead of going through a trustee bank issuer, as in the case of COPs. From there on the credit implications would be identical to those required for lease revenue bonds or COPs. These credit considerations would include the following:

1. A lease or lease purchase agreement with the lessee municipality of a term at least equal to the bonds issued by the EDA.
2. A pledge of the lessee municipality to make continuous lease rental payments over the life of the lease, or, if subject to annual appropriation, to supply a form of credit enhancement such as would be required for a COPs issue.
3. The assumption of all operation and maintenance expenses by the lessee, in addition to the payment of lease rentals and trustee fees.

Because both the issuer of the bonds (EDA) and the lessee (municipality) were bona fide political subdivisions, the leasing arrangement could be either a true lease or a lease purchase. Title to the parking facility and ultimate ownership would remain in the name of the tax-exempt entity in either case. Transfer of actual title from the EDA to the municipality could take place upon the retirement of the EDAs lease rental revenue bonds.

Public parking projects, meeting the restrictions of the 1986 Tax Law with respect to use and security and being financed by the issuance of economic development authority bonds, would *not* be classified as a private activity project, and the bonds issued would *not* be classified as private activity bonds. All of the limitations and prohibitions in the 1986 Tax Law concerning private activity bonds would not apply to the funding by an EDA or an IDA of such a public parking project owned and operated by a bona fide political subdivision. In effect, the local EDA is merely acting as the financing vehicle for the governing body of the municipality, given the authority to do so by state statutes.

Similarly, there is nothing to restrict an EDA from issuing the exact same bond issue to fund a parking project as a taxable obligation, if the use of the project or security behind the bond issue does not meet the tax exempt criteria of the 1986 Tax Law. The credit considerations and the bond security could be identical to a tax exempt bond issue, the only difference being in the rate of interest charged by the marketplace and the freedom from any arbitrage restrictions on the investment of bond proceeds. A recent example of such a concept was accomplished in New York State, where a series of tax exempt bonds and a series of taxable bonds (two public offerings) were issued for a single parking facility. The taxable series funded that portion of the garage being leased to private companies, and the tax exempt series funded the balance of the garage available to the general public.

In that example, the legal documentation, including the trust indenture and the lease agreement between the EDA and the lessee municipality, constituted one set of documents. Provision was made in the authorizing trust indenture for the issuance of bonds in two series, one taxable and one tax exempt. Because the municipality wanted to accede to the desires of the local private companies (utilities) with respect to the guarantee of employee parking through a long term lease of space and at the same time wanted to provide parking for a demonstrated public demand in the area, the compromise funding was devised of both taxable and tax exempt obligations for one project. At this point, it is well to remember the distinction between private, for-profit parking facility development and public-private parking projects developed basically at the request of a municipality. There is no way that the former can take advantage of tax-exempt financing in any of the methods outlined previously. However, the public/private parking facility could be financed on a tax-exempt basis, but only if it meets the restrictions in the 1986 Tax Law with respect to use, security and essential ownership. Obviously, tax-exempt financing limits the role of a private developer of a parking facility to that of a for-profit contractor up front and possibly a for-profit managerial role, over a short period of time. If a private developer is desirous of building a parking facility, which can be controlled for the use of tenants or users of commercial space, then the conventional means of financing must be used. If that private developer is willing to relinquish control of the parking use, open the facility to the public and eventually deed the facility to the host municipality, then several methods of tax-exempt financing would be available to him.

Tax Exempt COPs

The use of tax exempt certificates of participation (COPs) to finance new municipal parking projects contemplates the partial privatization of the proposed project. It is a form of long term financing that has been utilized for many years in connection with the municipal acquisition of equipment such as vehicles and computers. In recent years, however, the same concept has been extended successfully to include real property improvements. See Appendix 8 for a chart showing the structure of tax exempt COPs financing.

Tax exempt COPs financing provides funds for the construction of new parking facilities and all costs associated therewith. The basic documents include a lease-purchase agreement between a private sector lessor company and a municipal entity as lessee. Payments are made to the lessor (or to the trustee on behalf of the lessor) over a period of time by the lessee or buyer, while investors in the lease-purchase revenue instruments (COPs) provide the funds.

For interest on a bond or an obligation secured by a lease purchase agreement to be exempt from Federal income taxation, the issuer or the lessee must be a "State, territory, a possession of the United States, the District of Columbia, or any political subdivision thereof."

A political subdivision is defined as "any division of any State or local government unit which is a municipal corporation or which has been delegated the right to exercise part of the sovereign powers of the unit" (Treas. Reg.1.103-1). Three sovereign powers have been identified: the power to tax, the power of eminent domain, and the police power. It is not necessary that all three of these powers be delegated to the political subdivision. However, if only an insubstantial amount of any or all of such sovereign powers is delegated, then the Internal Revenue Service has stated that political subdivision status has not been achieved. Most municipalities qualify as political subdivisions, both as a "municipal corporation" and by possessing the requisite sovereign powers.

Structure of Tax Exempt COPs

The structure for accomplishing tax exempt COPs financing for a proposed parking facility would be as follows:

1. ABC Parking Company would be designated as developer, lessor, and nominal owner of the proposed parking facility.
2. The parking authority or the city would be the lessee of the proposed parking facility, as constructed.
3. A suitable bank trustee would be designated by the lessor and the lessee, to perform the following functions:
    - issue certificates of participation (COPs) in an amount sufficient to provide for all construction and associated costs for the proposed parking facility,
    - secure each certificate of participation by a proportional interest in the stream of basic lease rental payments made by the lessee, pursuant to a lease-purchase agreement,
    - hold in trust the proceeds from the sale of the COPs (to an underwriter), to be disbursed by invoice for the construction and associated costs of the proposed parking facility, and
    - to receive the basic lease rental payments from the parking authority or the city and disburse same to the holders of the COPs, for and on behalf of the lessor,
4. The city and ABC would enter into a suitable land lease for a parcel of land of a size consistent with the requirements of the proposed parking facility, at a nominal annual lease rental.

5. The lessor (ABC) would covenant in the lease-purchase agreement to bear full responsibility for the construction and placing in operation of the proposed parking facility, pursuant to plans and specifications previously approved by the parking authority or the city, as lessee; it is recommended that a guaranteed maximum price and completion date be set by the lessor.
6. Title to the completed parking facility would be in the name of the parking authority or the city, as lessee, held by the trustee bank subject to the lien of the lessor for the prompt payment of annual basic lease rentals.
7. The obligations of the parking authority or the city, as lessee, pursuant to the lease-purchase agreement would include:
   - the payment of a basic lease rental equal to the annual principal and interest requirements of the COPs issued to finance the total costs of construction and associated costs of the proposed parking facility,
   - assume the obligation and all costs in connection with the operation and maintenance of the parking facility,
   - a covenant to operate the parking facility as a public parking facility, in accordance with applicable Federal laws,
   - a covenant to annually budget and appropriate from the general fund sufficient moneys to make the basic lease rental payment of the city or the parking authority.

Based on the provisions of the lease-purchase agreement, with respect to the city's or parking authority's pledge to meet the required lease rental payments, tax exempt COPs financing as described could achieve an investment grade rating and be eminently marketable at reasonable rates. Inasmuch as the term of the lease-purchase agreement is limited to the useful life of the financed facility, determined by current statutes, the term of any tax exempt COPs financing cannot exceed a like term. Within that limitation, however, a great deal of flexibility can be achieved in the structuring of the financing, without the restrictions associated with general obligation bonds.

CONSTRUCTION RISK

In the scenario previously described, the lessor assumes the obligation of construction of the parking facility, albeit protected to a degree by a maximum guaranteed construction price. In general, however, investors in tax-exempt instruments are risk averse and may not be willing to invest in a project when repayment of their investment is conditioned upon successful completion of construction and the resulting commencement of lease payments. Pursuant to the lease-purchase agreement, lease rental payments from the parking authority or the city could commence upon the issuance of the tax exempt COPs or on a specific date thereafter and not be contingent upon the completion of the facility. In this scenario, the construction risk would be borne by the parking authority or the city, as lessee. By removing this construction risk from the investor, the interest rate borne by the tax exempt COPs would approach those for city general obligation bonds. The most successful lease-purchase financing, therefore, requires the governmental body to either make lease rental payments in the event the parking project is not completed, or to obtain insurance against construction and completion risk by a third party, so that neither the government body, the tax exempt COPs holders nor the lessor bears this risk. In some instances, where the lessor is a nominal corporation created solely for this purpose, and not an experienced parking company, the lessor designates the municipal lessee as its agent for the construction and completion of the proposed facility.

DEPRECIATION

For Federal and state income tax purposes, a lease purchase transaction would be considered a conditional sale implying ownership by the government body. For this reason, no tax benefits, such as depreciation or investment tax credit would be available to any private tax-paying developer involved in the financing. Conversely, the improvements would not normally be subject to real estate taxes, as the governmental body is the user.

ANNUAL APPROPRIATION

In the initial tax exempt COPs scenario previously described, the obligation of the parking authority or the city, as lessee, to make lease payments was described as a general obligation payable directly or indirectly from the general fund of the city. This was meant to imply a continuous obligation over the life of the lease. Under such an obligation, the amount of the tax exempt COPs financing would probably be held, by bond counsel, to be a debt of the city and includable in its debt statement. However, the lease-purchase agreements for many tax exempt COPs financings are structured such that the lease rental payment is subject to annual appropriation by the governing body. The lease and the subsequent financing (COPs) provided by the lessor in such cases is not considered a debt obligation of the governmental body, if the lease payments are subject to annual appropriation. Such a distinction would not necessarily apply to a parking authority, which would not be subject to statutory debt limits.

For this reason, lease-purchase financing is often used in situations where governments are constrained by debt limitations, structure restrictions, or referendum requirements regarding the issuance of debt. The rating ascribed to a tax exempt COPs issue, however, reflects the ability of the lessee to release his obligation in the event of non-appropriation, and the possibility that the transaction structure will collapse in such a circumstance. Typical non-appropriation language included in the lease, would be similar to the following:

> "Lessee reasonably believes sufficient funds can be obtained to make all lease payments due during the term of the lease. Lessee hereby covenants that it will do all things lawfully within its power to obtain and maintain funds from which lease payments may be made, including making provisions for such payments to the extent necessary in each biannual or annual budget submitted for the purpose of obtaining funding, using its bona fide efforts to have such portion of the budget approved, and exhausting all available reviews and appeals in the event such portion of the budget is not approved."
>
> "In the event that the lessee is unable to obtain budget funding for any renewal term, lessee shall have the right to terminate this lease at the conclusion of the then current term of the lease and shall neither be obligated to make any lease payments due beyond the current term, nor to make any concluding payment whatsoever, and this lease shall terminate as to that leased facility. Provided, however, that in the event lessee does not appropriate such funds, lessee will use its good faith best efforts to acquire the necessary funding from other agencies or sources. Upon termination as provided for above, lessor or its agents shall have the right to take possession of the leased equipment/facility and lessee shall be liable to return the leased equipment/facility to lessor in full operational and good working order."

Due to these provisions for non-appropriation, the security provided to the tax exempt COPs holders in the event that the government unit ceases to make lease payments, and therefore provide for certificate debt service, is a critical element to the rating of tax exempt COPs. The most fundamental issue will be the ability to grant a security interest, or, in other words, give the certificate holders the right to the ownership of the facility financed, in the event payments are no longer made. The ability to grant such an interest is not universal, and in certain states this security feature cannot be included. In those states where a security interest is not a legal remedy, the need to develop strong alternative remedies for the certificate holders is a major concern. One such alternative is to provide to the certificate holders the proceeds of a subsequent sale of the financed facility. Increased use of tax-exempt leases has provided a number of examples of how to structure an adequate remedy in those states where a security interest is not available.

INDEMNIFICATION

In a lease financing, the lessor is a party to the critical agreement in the structure—the lease-purchase agreement. Although the lessor will typically transfer all rights and obligations to the trustee, a claim against the lessor could conceivably collapse the transaction. In order to minimize the likelihood of such an occurrence, the municipality will typically indemnify the lessor against all claims relating to the use or ownership of the facility. The need for such a clause relates to the possibility of a liability suit filed against the named lessor relating to improper use of the leased facility, or a tax levied at some point in time on lease payments made to the lessor. In certain states, the lessee has not been able to grant such an indemnification. In those cases, a private corporation is effectively precluded from acting as lessor, and a state or local agency is alternatively used as the nominal lessor.

INTERIM FINANCING

A part of the process of designing and constructing a new parking project consists of the following work which is done usually prior to the issuance of long term bonds to fund the project:

- survey of alternative sites
- demand and feasibility studies
- site borings and sub-soil tests

- land acquisition
- preliminary design
- final design to working drawings
- preparation of construction bidding specifications
- preliminary legal fees
- building permit fees

Considerable care should be taken by the authority or municipality proposing the construction of a new parking project, to keep track accurately of such preliminary expenses attributable to the project. Regardless of the source of funds used to meet these expenses, if they are to be reimbursed from the proceeds of a tax exempt bond issue, an accurate record must be kept, for proper authentification by the trustee for disbursement of such expenses from a funded construction account.

Several options are available to a municipality to raise interim funds to meet the previously mentioned preliminary costs. The usual sources are as follows:

1. periodic appropriation from the general fund of a municipality or the capital improvement fund of an authority,
2. short term borrowing from a local bank,
3. the issuance of bond anticipation notes,
4. the use of available Federal or state grants.

Appropriations from a municipality's general fund, periodically over time as such preliminary expenses are incurred, is the most prevalent method used. Sometimes where an authority is involved as the developer of the parking project, the authority will temporarily borrow such general fund appropriations from the municipality. If the authority has sufficient funds in a capital improvement account, such borrowing from the municipality would not be necessary. The use of general fund appropriations to fund these preliminary expenses not only eliminates the cost of borrowing, but, more importantly gives the municipality the opportunity to assess the project constantly, during the development phase. If for any reason, particularly economic feasibility, the governing body of the municipality decides to cancel the project, it can be done with no further obligation. Even a decision to put the project on hold, pending improved economic developments or greater parking demand, can be done. The preliminary studies and engineering work can be held for future updating and use.

If general funds are not immediately available for this purpose, a municipality or an authority usually has the ability of borrowing short term funds from a local commercial bank. It is not uncommon for the local bank holding the normal accounts of the municipality, to make such short term loans available. For the purpose of meeting periodic preliminary project expenses, as described, such a short term bank borrowing should take the form of a line of credit with the bank, specifically for the purpose of the parking project. In such an instance, the line of credit would be secured by the general obligation of the municipality or the authority, provided the latter's credit standing was adequate. In effect, the bank would be loaning the amount required based on the general credit of the borrower, not on the economics of the proposed parking project.

The usual practice of structuring such a line of credit, once the line is opened by the bank, is for the borrower to draw down such amounts as are needed for the specific purpose, usually on a monthly basis. An estimate of a draw down schedule is presented in the beginning. The term of the total loan can be any reasonable period acceptable to the bank, usually not more than one year. Repayment of moneys borrowed can be made from the proceeds of permanent financing or, if delays are incurred, the loan can usually be rolled over for a new term. There should be, included in the loan documents, the right of the municipality to repay the balance of the loan outstanding, with accrued interest, at any time.

It is also usual practice for such a bank to charge a monthly commitment fee for the aggregate amount of that portion of the total loan not yet drawn down. This compensates the bank for the reservation of the funds subject to draw down. Upon drawing down a portion of the line of credit or reserved funds, the bank then charges the negotiated interest rate on the aggregated amounts drawn down.

Up-front or "issuance" costs related to the bank borrowing, which would have to be paid by the municipality, but which could be reimbursed from the proceeds of such a line of credit would include the following:

Description	*Estimated Charges*
In-house counsel, document preparation	$ 5,000
Bond counsel, tax-exempt opinion	5,000
Bank legal fees	10,000
Total	$20,000

In an effort to quanify the amount of interim financing required, assume that an expenditure of $700,000 for project preliminary expenses will be required and that such funds will be spent at the rate of $100,000 per month over a period of seven months to the completion of all engineering and other preliminary expenses; an additional $20,000 has been included to cover the issuance costs in the first month.

Table 7-1 shows an estimated drawdown of a maximum $720,000 line of credit, with the monthly commitment fee and interest charges over a one year period. Discussions with several commercial banks have indicated that a municipality could not expect to benefit from its tax-exempt status, in the rate of interest charged by banks for a line of credit. Two reasons were given: (1) the relatively small amount of the anticipated line of credit, and (2) the fact that most banks today cannot use tax exempt income, to the extent they could prior to the 1986 Tax Law. For these reasons, the following schedule projects the interest charges for the line of credit as being at 9.0%.

Bond Anticipation Notes

Bond anticipation notes (BANs) are issued through the authority of the basic bond resolution giving the issuer the authority to issue long term bonds. A part of the bond resolution permits the issuer to emit notes "in anticipation" of the issuance of bonds. The requirements include a limitation on the term of the notes, usually one year, and a

Table 7-1. Estimated Line of Credit Costs. $720,000 Maximum

MONTH	ESTIMATED MONTHLY DRAW-DOWN	REMAINING LINE OF CREDIT	½ OF 1% COMMITMENT FEE	AGGREGATE BORROWED AMOUNT	9.0% INTEREST CHARGES	TOTAL COSTS
1	$120,000	$600,000	$250	$120,000	$ 900	$ 1,150
2	100,000	500,000	208	220,000	1,650	1,858
3	100,000	400,000	167	320,000	2,400	2,567
4	100,000	300,000	125	420,000	3,150	3,275
5	100,000	200,000	83	520,000	3,900	3,983
6	100,000	100,000	42	620,000	4,650	4,692
7	100,000	—	—	720,000	5,400	5,400
8	—	—	—	720,000	5,400	5,400
9	—	—	—	720,000	5,400	5,400
10	—	—	—	720,000	5,400	5,400
11	—	—	—	720,000	5,400	5,400
12	—	—	—	720,000	5,400	5,400
Total	$720,000	—	$875	$720,000	$49,050	$49,925

provision that the notes *may* be paid from the proceeds of the sale of bonds issued pursuant to the bond resolution. One or more "roll-overs" are permitted, whereby a new issue of BANs may be issued to retire a maturing issue. Some states limit the period of time for BANs to be rolled-over, so care must be taken that the requirements of law are observed.

The sale of BANs will require the issuance of an unqualified legal opinion as to Federal tax exemption, by bond counsel, whether the BANs are marketed in a public offering or as a private placement to institutional investors. The documentary requirements would be comparable, regardless of the method of sale, except that more expense would be incurred for the publication of a full official statement in the case of a public offering. A private placement could be accomplished with an in-house published offering memorandum.

The disclosure required would include the following:

1. a full description of the municipality,
2. financial statements of the municipality over the last several years and the current budget,
3. a full description of the project being financed, with cost estimates and time schedule,
4. description of the engineers (design) and any construction contractors involved,
5. a full economic feasibility study, indicating parking demand, rates, projected annual revenues and expenses over at least 10 years and estimated coverage of annual bond debt service for the full amount of the permanent financing authorized,
6. a summary or full copy of the authorizing bond resolution or trust indenture,
7. a copy of the proposed legal opinion of bond counsel
8. a description of the terms of the notes being offered, with the names of the trustee and/or bank paying agent.

The issuance costs for a municipality, in connection with BANs, would include the following:

1. *Bond Counsel*—because all of the basic legal work normally to be done for the issuance of bonds, would have to be done prior to the issuance of BANs, the expenses for bond counsel would be virtually the same as for the issuance of bonds; the amount is subject to negotiation, but could range from $50,000 to $75,000 or more, depending on the size and complexity of the financing.

2. *Rating Agencies*—application for a rating from Moody's Investors Service and Standard & Poor's Corporation is recommended; the cost for these ratings will aggregate about $15,000, depending on the complexity of the proposed financing.
3. *Publication Costs*—the printing of an official statement, which would be required for a public offering of BANs, would cost an estimated $25,000, including handling and postage; the cost of an in-house published offering memorandum required for a private placement can be estimated at not more than $5,000, including reproduction and mailing.
4. *Accounting Costs*—if interim stub figures are required from the municipality's accountants (which depends on the timing of the offering vs. the date of the last audit report), some expenses would be incurred for this purpose; probably not more than $10,000 would be required.
5. *In-House Legal Expenses*—the cost for the in-house supervision of the transaction and closing by the municipality's attorneys could be estimated at not exceeding $20,000.
6. *Trustee's Fees*—the acceptance fee and legal expenses of the trustee bank for the BANs (and for the permanent financing) can be estimated at not to exceed $10,000.

A summary of the anticipated issuance costs in connection with the issuance of BANs would be as follows: (maximums)

Description	*Public Offering*[1]	*Private Placement*[1]
Bond counsel	$ 75,000	$ 75,000
Rating agencies	15,000	15,000
Publication costs	25,000	5,000
Accounting costs	10,000	10,000
In-house legal expenses	20,000	20,000
Trustee's fees	10,000	10,000
Total	$155,000	$135,000

[1]Issue size not less than $5 million.

These estimated issuance costs can be included in the sizing of the amount of BANs offered and paid from the proceeds of the sale thereof. A substantial portion of these issuance costs, such as bond counsel expenses and rating agency fees, will *not* be duplicated for the issuance of bonds by the municipality for the permanent financing of the project.

Federal or state transportation grants are sometimes available for the planning and development process for a new parking project. The availability of such funds is dependent upon the following several factors:

- the total level of such funds available from Federal or state appropriations,
- the granting criteria used for the allocation of available moneys,
- the eligibility of the proposed parking project within the granting criteria, and
- the limitations or restrictions for which the grant moneys may be used.

Frequently, such grants as may be available are conditioned upon the money being used in urban renewal areas or economically depressed areas. In addition, sometimes such grants may only be used for tenant relocation, environmental impact studies, overall transportation planning, or public transit planning. However, it is worthwhile for a municipality to investigate and apply for any such grant moneys, in order to reduce, if possible, the overall cost of planning and constructing a new parking facility.

MUNICIPAL BOND DEFAULTS

Tax exempt municipal bonds, as a class of securities, are second only to obligations of the United States government, with respect to the safety record of the payment of interest and principal to investors. Analysts of the subject opine that there are too many factors that effect the credit of individual bond issues or even of whole categories of municipal bond issues to depict accurately the risk of default. In addition, they claim that the rapidly changing markets of the 1980s have rendered much of the historical data nearly worthless in comparison.

According to information gathered at the University of Rochester, the total number of defaults between 1980 and 1988 (about 600), represented 0.72% of the 83,000 municipalities counted by the Census Bureau. The $582.9 million of municipal debt of all kinds that went into default over the nine years amounts to 0.81% of the total tax exempt municipal debt outstanding.

Municipal research analysts agree that the most likely type of municipal bond to default is an industrial development bond, secured by a private company. By far the majority of such defaults over the past 40 years have been IDBs.

CHAPTER 8

RATE STRUCTURES FOR TAX EXEMPT OBLIGATIONS

The vast majority of new issue tax exempt project financing by municipalities consists of fixed rate obligations with serial maturities or a combination of serial maturities and term maturities with a full sinking fund. In a few states, statutes dictate the term over which a new project may be financed, the latest date, after issuance, when the first amortization or principal payment must be made and the minimum and maximum proportional amounts which must be amortized annually.

All states permit municipalities to issue bonds bearing a fixed interest rate for each principal maturity. For a full serial bond issue, this could and frequently does result in a series of fixed interest rates, one for each year of the serial maturity schedule.

Fixed interest rates are the most conservative form of borrowing by municipalities. However, the timing of capital needs does not always coincide with the best market conditions for the sale of such bonds. During periods of high interest rates, municipalities are sometimes forced into borrowing at long term fixed rates which may be historically higher than they are used to paying. Where redemption features are included in such bond issues, municipalities can return to the market during periods of lower interest rates, to refund outstanding high rate issues and effect borrowing savings. See Appendix 10 for a chart of comparative interest rates, since 1983.

VARIABLE RATE OBLIGATIONS

Variable rate borrowing provides an alternative to the standard fixed rate, level debt service, serial or term maturity financings used historically. It permits more flexibility in the timing of funding capital requirements, it is more responsive to market movements, and it can be converted more quickly to other modes or maturity structures. On the other hand, it is less predictable than fixed rate borrowing, and, in periods of inverted yield curves, it can be more costly. See Appendix 11 for a chart of historical rates of short term, tax exempt obligations, as measured by the Kenny Index.

Prior to 1973, variable rate borrowing, other than short-term rates with commercial banks, was virtually unknown to the tax-exempt marketplace. All forms of short-term variable rate borrowing had been well utilized by the taxable corporate market, but municipalities were slow to adapt these forms of borrowing to their own needs, largely because the volume of long-term fixed rate tax-exempt bonds and their interest rates were so low that there was no need on the part of municipal borrowers to assume the volatility risk that accompanies the benefit of variable rate borrowing. As the volume of municipal borrowings grew and the competition for tax-exempt buyers increased, rates began to rise—in response to supply/demand factors. In 1973, tax-exempt commercial paper (TECP) was created in response to a request from corporate clients familiar with commercial paper in the taxable markets. TECP enabled municipalities to take advantage of the lower end of the tax-exempt yield curve when long-term rates were very high. Since that time, many other short-term financing modes were created: weekly variable rate demand bonds (VRDBs) gained popularity and became the most popular short-term, tax-exempt financing mode since they were created in 1973; money market municipals (MMM) provided a substitute for TECP in 1984, when impending tax law changes threatened to impose restrictions on the municipal markets; and in 1988 select auction variable rate securities (SAVRS) eliminated the need for liquidity support as the cost of such support increased.

The salient features of these four variable rate financing modes are described as follows:

Tax-Exempt Commercial Paper (TECP): TECP is a short-term unsecured promissory note that exhibits the following characteristics:

- *Short-term Maturity:* Maturities usually average under 90 days but may be as short as 1 day or as long as 365 days.
- *Lowest Interest Rates:* TECP and its long-term equivalent, MMMs, provide the lowest cost of funds of all the variable rate options.
- *Arbitrage Rebate Management:* TECP provides issuers with the ability to avoid arbitrage rebate by spending proceeds within the six month temporary period, permitted in the 1986 Tax Law. For a construction project, just enough TECP should be issued to satisfy six months of expenditures at a time. This enables issuers

to take advantage of temporary period regulations, earn arbitrage for as long as six months and avoid arbitrage rebate. If proceeds remain after six months, rebate will be necessary.

- *Flexible Amounts Outstanding:* TECP works well to provide for any short-term cash flow deficiencies or capital borrowings with an expectation of periodic prepayment. TECP should be issued in amounts as needed and paid down from money received when the TECP matures. The total amount borrowed would not need to be outstanding for the life of the obligation, thus reducing the amount of debt service payable.
- *Reissuance:* If more funds are needed as the TECP matures, a reissuance occurs and is subject to changes in tax laws in effect at the time of each reissuance. Adverse changes in the tax code would only occur after a great deal of publicity, however, providing advance warning and enabling issuers to refund TECP with some other short-term security.
- *Cost of Issuance:* TECP has lower initial issuance expenses because of the absence of an underwriting discount, making it the least expensive technique to use when a need exists requiring an increase in the amount borrowed. Ongoing expenses include a remarketing fee and liquidity letter of credit fees.
- *Credit and Liquidity Support:* TECP issues need only short-term ratings. The market for TECP will accept only those issues bearing the highest short-term ratings, P-1 from Moody's and A-1 or A-1+ from Standard and Poor's Corporation, so liquidity support will be needed unless an issuer carries such ratings. The liquidity support, traditionally a letter of credit secured from a commercial bank for TECP issues, is to assure the timely payment of principal and interest on maturing TECP.

Variable Rate Demand Bonds (VRDBs): VRDBs are a long-term financing structure with a periodic repricing mechanism. VRDBs exhibit the following characteristics:

- *Floating Rates:* VRDBs interest rates change at regular intervals, usually weekly, but can be set to monthly, quarterly, semiannual, annual, biennial, or three-year repricings.
- *Periodic Remarketing:* The remarketing agent reprices the bonds to determine the new rates from one period to the next period.
- *Single Issuance:* VRDBs provide a single issuance of long-term, tax-exempt debt obligations under current law and regulations,

and therefore are subject to laws in existence at the time of issuance and avoid the "reissuance" concern.

- *Par Valuation:* VRDBs value equals par during the floating rate period.
- *Conversion Options:* VRDBs retain the option of converting to another VRDB mode, MMM, SAVRS, or a fixed rate mode at such time as the conversion fits the issuer's needs.
- *Low Interest Rates:* Weekly VRDBs provide a low cost of long term borrowing, as shown in the interest rate comparison below. The remarketing agent prices a VRDB issue at the highest incremental rate necessary to remarket the entire issue to all investors, however, so the weekly VRDB usually costs slightly more than TECP or MMMs priced at the same time.
- *Uses of VRDBs:* The VRDB structure may be considered as a permanent part of an issuer's debt management plan or as a form of interim financing with the mechanics and documentation for an eventual long-term, fixed rate financing already in place. If used as a form of interim financing, VRDBs can lower the costs of capitalized interest during construction or provide an alternative to fixed rates during a high interest rate environment.
- *Low Minimum Size:* The market does not impose a minimum size requirement for VRDBs, because size is not required to maintain liquidity for remarketing purposes.
- *Credit and Liquidity Support:* VRDB issues must have both short and long-term ratings. Like the TECP market, the market for VRDBs will accept only those issues bearing the highest short-term ratings, VMIG1/A-1 or A-1+, so credit support may be needed unless an issuer carries such ratings. Issuers also would need a long-term rating of A+ or above to issue debt without credit enhancement. Liquidity support, a letter of credit generally secured from a commercial bank regardless of long-term ratings, is to assure the timely payment of principal and interest on maturing VRDBs.
- *Arbitrage Rebate Management:* Unlike TECP, VRDBs are subject to arbitrage rebate requirements if funds are not spent within the first six months after issuance of the bonds.

Money Market Municipals (MMMs): MMMs combine investor preferences for low interest rates offered by TECP with the long-term VRDB structure to achieve a single issuance of debt for tax purposes. MMMs possess the following characteristics:

- *Short-term Rates:* MMMs interest rates must be reset at short-term money market rates for periods ranging from 1 to 365 days, in a manner similar to TECP, to achieve the lowest rates in the municipal marketplace for long-term bonds.
- *Low Remarketing Risk:* The MMM structure reduces the remarketing risk of the issue, because each MMM bond is remarketed in small amounts on prescheduled tender dates, whereas the entire VRDB issue may require remarketing on any optional tender date.
- *Single Issuance:* MMMs should satisfy any bond counsel's reissuance concerns, because the repricing discretion is limited solely to the professional judgment of the remarketing agent.
- *Par Valuation:* MMMs value always equals par during the floating rate period.
- *Conversion Options:* MMMs provide the issuer with the option of converting to a variable, SAVRS or fixed rate at such time as the conversion fits the issuer's needs.
- *Low Interest Rates:* The interest rates on MMMs, along with TECP, are the lowest of the variable rate obligations, because each MMM bond is priced at rates customized for each investor in small amounts, typically averaging $2,000,000.
- *Uses of MMMs:* The MMM structure can be used as a permanent part of an issuer's debt management program or as a form of interim financing with the mechanics and documentation for an eventual long-term, fixed rate financing. MMMs, as a form of interim financing, may be beneficial to avoid high rates during a high interest rate environment or during construction to lower the cost of capitalized interest.
- *$20 Million Minimum Size:* The remarketing agent prefers a $20 million minimum size to increase the ability to sell different sized pieces when remarketing.
- *Credit and Liquidity Support:* MMM issues must carry both short and long-term ratings. Like the markets for TECP and VRDBs, the market for MMMs will accept only those issues bearing the highest short-term ratings, VMIG-1/A-1 or A-1+, so credit support may be needed unless an issuer carries such ratings. Issuers also would need a long-term rating of A+ or above to issue debt without credit enhancement. Liquidity support generally secured from a commercial bank regardless of long-term ratings, is to assure the timely payment of principal and interest on maturing MMMs.

- *Arbitrage Rebate Management:* Unlike TECP and like VRDBs, MMMs are subject to arbitrage rebate requirements if funds are not spent within the first six months after issuance of the bonds.

Select Auction Variable Rate Securities (SAVRS): SAVRS are a long-term financing structure with short term rates priced by the Dutch Auction process. SAVRS exhibit the following characteristics:

- *Periodic Repricing:* The SAVRS interest rate period lasts for 35 days, then is reset through a Dutch Auction pricing mechanism. In a Dutch Auction, potential or existing holders may submit orders to hold, sell or bid for bonds to the Auction Agent. The winning bid rate is the rate that, in combination with all lower bid rates, clears the market of the entire principal amount of SAVRS available for sale and becomes the interest rate on all bonds for the next auction period.
- *Elimination of a Liquidity Facility:* The Dutch Auction process removes the contingent obligation of an issuer to purchase investor's bonds at par on any given interest reset date, thus eliminates the need for liquidity support and associated costs.
- *Uses of SAVRS:* SAVRS may be used for the same purposes as VRDBs and MMMs, as a permanent part of an issuer's debt management program or as a form of interim financing with the mechanics and documentation of an eventual long-term fixed rate financing already in place. As interim financing, SAVRS may lower the cost of capitalized interest during construction or may provide an alternative to fixed interest rates during a period of high interest rates.
- *$40 Million Minimum Size:* The market prefers SAVRS transactions in amounts equal to at least $40 million for liquidity. As SAVRS become more popular, the market may accept smaller issues.
- *Par Valuation:* SAVRS value always equals par on the auction date.
- *Conversion Options:* SAVRS provide the issuer with the option of converting to a variable, MMM, or fixed rate at such time as the conversion fits the issuer's needs.
- *Low Interest Rates:* SAVRS are priced competitively relative to other forms of short-term debt, because SAVRS do not need liquidity support.

- *Long-Term Debt Treatment for Accounting Purposes:* Because SAVRS do not rely on underlying third party liquidity arrangements, the rating agencies and accounting concerns classify SAVRS as long-term debt obligations with a long-term bond rating.
- *Credit Support:* SAVRS issues require only a long-term rating. The market for SAVRS will only accept those issues bearing an A rating or better. Issuers with a long-term rating below an A, would need to seek credit enhancement for the SAVRS issue.

COMPARISON OF VARIABLE RATE OBLIGATIONS AND FIXED RATE NOTES

Table 8-1 summarizes the comparison of some of the main features of variable rate obligations the average interest rate is as of February 11, 1991. Each financing structure has an appropriate use and function depending upon an issuer's needs.

Table 8-1. Summary Comparison

	VARIABLE RATE PROGRAMS			
ITEM	TECP	VRDBs	MMMs	SAVRS
Average Interest Rate	4.00%	4.00%	4.00%	4.50%
Initial Issuance Costs[1]	0.36	0.61	0.63	1.04
Ongoing Cost[1]	0.425	0.425	0.425	0.30
Liquidity Facility	Yes	Yes	Yes	No
Reissuance	Yes	No	No	No

[1]A percentage of the par value issued, as a one-time cost.

CHAPTER 9

THE IMPACT OF RATINGS ON TAX EXEMPT PARKING BONDS

CREDIT RATINGS

The credit rating assigned to a debt security is the expression of the opinion of the rating agency of the relative degree of probability of the timely repayment of principal and interest. With respect to tax-exempt municipal bonds, three such rating agencies currently analyze and assign ratings to new and outstanding bonds. Each of these rating agencies charges a fee for their rating work, sufficient to reimburse them for the time and expense of their analysis.

THE RATING AGENCIES

The two major agencies whose ratings are most widely accepted are Moody's Investors Service, a wholly owned subsidiary of Dun & Bradstreet, and Standard & Poor's Corporation, a wholly owned subsidiary of McGraw Hill. The third agency, Fitch Investors Service, does not rate as many municipal bonds and is not as widely recognized as the other two. All three agencies are headquartered in New York City. All three also publish various reports, studies, and periodicals as a normal part of the dissemination of information to their subscribers. When the agencies conclude their analysis of a new bond issue and assign a rating, it is published in each company's weekly publication, as well as being announced to other municipal bond trade publications, such as "The Bond Buyer" and the "Munifacts" broad tape system.

RATING ANALYSIS

With respect to tax-exempt parking bonds issued by municipalities or parking authorities, one of the most important sources of information utilized in the analysis of such bonds is the economic feasibility study prepared by independent professional engineers with a good track record in parking projects. Why is this so important, particularly if the parking bonds are general obligation bonds of a municipality? Some of the reasons are as follows:

1. The analyst can ascertain, from an unbiased source, the need for the project, the projected impact on traffic in the area, the anticipated production of revenues as compared to operating expenses and debt service charges and the impact, if any, of the proposed new parking debt on the overall debt structure of and tax burden on the municipality.
2. The analyst can determine the existence, if any, of other competing parking facilities, including their proximity and effect on parking traffic.
3. The analyst can examine the rate structure to be charged and, by comparison with other facilities in the municipality or the region, determine if the proposed facility can be expected to attract parkers.

A second, but equally important factor in the analysis of parking bonds is the nature of the security behind the bonds. It is in this area where the analyst must determine, (1) if the bonds are payable solely from parking revenues, (2) if they bear some sort of guarantee from the municipality, (3) if they are also general obligations of the municipality payable from unlimited property taxes, or (4) if they are secured by lease rental payments from a hospital, a municipality or some other entity. If the bonds are general obligations of the municipality, the terms and conditions of the authorizing bond resolution are fairly simple and straightforward, i.e., principal and interest on the bonds will be payable from unlimited ad valorem property taxes through the general fund of the municipality. However, if the bonds are anything less than a general obligation, then the detailed terms and provisions of the authorizing bond resolution become a major factor of analysis.

Some of the provisions in a revenue or lease rental revenue parking bond, which are examined carefully by the rating analyst, are as follows:

1. A clear cut "flow of funds," i.e., adequate language providing for the collection and deposit of revenues or lease rentals and the

disposition of same in a clearly defined order of priority, beginning with operating and maintenance expenses, through payment of debt service and reserves to surplus funds.

2. An "adequate rate covenant," wherein the issuer pledges to review and adjust rates for parking or other services provided in the parking facility, from time to time, as may be needed so that the revenues produced are not less than the aggregate of operating and maintenance charges, debt service and reserves.
3. A pledge to accumulate and maintain reserves for operating expenses, for debt service and for renewals and replacements, the latter in such amounts as may be recommended by independent engineers.
4. A limitation on the issuance of additional bonds in the future which would be payable prior to or on a parity with the proposed bonds; traditionally a test for additional parity bonds would be that the maximum annual principal and interest charges on all outstanding parking bonds, plus the proposed additional bonds, would be covered by available net revenues from existing parking system facilities in each of the preceding three years at least 1.30 times, plus being covered by projected net revenues, including the new parking facility being financed, at least 1.50 times in each year over the life of the new bond issue.

These points are just a few of those items in the authorizing bond resolution which must be studied and evaluated by the rating analyst. The relative strengths and weaknesses of these various provisions as the analyst applies his/her expertise as if he/she were a potential investor, results in an overall appraisal of the security aspect of the bond resolution.

In addition to the economic feasibility of the project and the degree of security provided by the bond resolution, the rating analyst must broaden his investigation to include the community as a whole. Parking facilities in particular are dependent upon changing economic conditions in the local area which they serve and in the community as a whole. By examining various statistical indicators, such as building permits, assessed valuations, retail sales, population trends, etc., the analyst can determine if the area economy is healthy, stagnant, or declining. Debt statistics, tax collection records, municipal budget performance and other information will indicate to the rating analyst the ability of the municipal administration to manage its finances. In view of the somewhat less than satisfactory performance in this area by some of our larger cities in recent years, this factor of the analysis has become increasingly important. In the event that the proposed parking bonds are general obligations of or guaranteed by the municipality, such detailed statistics and other information about the community would be necessary for the analyst to assess the impact of such debt upon the community.

PROBLEMS WITH PARKING REVENUE BONDS

In considering the analysis of parking bonds payable solely from the revenues of one or more parking facilities, the rating analyst must take into account several factors which would not be applicable to those bonds bearing the additional security of the unlimited taxing power of the issuer. Of primary and direct importance, relative to the ability of one or more parking facilities to produce sufficient revenues to meet annual debt service on parking revenue bonds, is the historical record of those revenues produced by any existing facilities pledged under the proposed financing, as well as the revenues which may be projected from new facilities to be financed.

It is extremely rare to find a parking authority that has the enviable ability to finance a new parking facility based on a demonstrated "coverage" of annual debt service, including the proposed new financing, from historical net revenues of existing facilities alone, without giving any effect to the revenues to be derived from the new facility to be financed. As a consequence, the importance of a professional economic feasibility report prepared by qualified engineers specializing in the parking industry, cannot be overemphasized. The depth of research and analysis done by such an engineering company in the feasibility report will be studied carefully by the rating analyst, as well as potential investors, and the credibility of the results will depend on the on-site research and bases of assumptions made by the engineers. If the assumptions of traffic, turn-over, parking rates, or operating expenses appear unreasonable, the credibility of the entire report will be lost.

However, because of the inherent vulnerability of a single parking facility to meet, over a long period of time, projected net revenues equal to the retirement schedule of a bond issue, it is also rare to find a high rating applied (or any rating at all) to a bond issue supported by the revenues of a single parking facility. This axiom will hold true unless the feasibility report demonstrates the existence of an adjacent traffic generator of such a magnitude and continuity that a debt service coverage of close to 2.00 times is shown.

The obvious answer to the problem of maximizing the credit worthiness (rating) of parking revenue bonds is to include in the pledge of available revenues the operations of multiple parking facilities, i.e., the parking system approach. By including in the revenues available to meet debt service on parking revenue bonds, the revenues from *all* of the parking facilities under the control of a parking authority or parking commission, a greater degree of parking diversification and a broader base of parking revenues is obtained. Thus, the major detriment to single facility financing is ameliorated, if not eliminated.

In addition, the inclusion of a pledge of on-street parking meter revenues adds to the base of user revenues available to meet debt service. In some communities, such meter revenues are set aside in a special fund, to be drawn down by the parking agency only if needed to meet a deficiency from other off-street parking revenue sources. At the end of each fiscal year, if such meter revenues are not utilized, they flow into the community's general fund. If clearly stated, such a pledge of meter revenues is acceptable to the rating analyst.

There are several other specific points of potential vulnerability which are inherent in the analysis of parking revenue bonds. The rating analyst must not only be aware of these points but must be satisfied as to their degree of importance before reaching a conclusion as to the rating to be assigned to such bonds. Without attempting to specify the priority of these potential areas of vulnerability, the following are of concern to the analyst:

1. In the subject community, what degree of competition is provided by private parking operators to the municipal parking agency? Does the municipal parking agency own or control most, if not all, of the prime downtown parking sites or is the municipal ownership of parking facilities in the minority, with a strong private ownership bloc well represented at "City Hall"?
2. Are the existing or proposed new parking facilities of the public agency in prime locations which are adjacent to major vehicular traffic generators, such as hospitals, downtown medical clinics, universities or colleges, downtown shopping malls, major hotels

or motels, sports coliseums or arenas, theatres or government buildings? To what degree is a single parking facility dependent on a single source of traffic generation and what type of traffic will be generated (monthly, transient, daytime, or evenings only)?

3. Most public parking agencies and their attorneys are aware of the provisions of the Federal Clean Air Act, as recently amended and as promulgated by the Environmental Protection Agency and each state EPA agency. The analysts at the rating agencies and many investors are also aware of the vulnerability of every public parking agency in the United States to the awesome power of the EPA with respect to the potential shut-down of inner-city parking facilities under certain conditions of air pollution. The impact of such a potential shut-down and the temporary cessation of revenues must be weighed by the analyst. The existence of business interruption insurance, if available, would constitute an off-setting factor.

RATINGS VS. INTEREST RATE

It is already obvious that for municipal bond issues of all kinds, the higher the rating the lower the rate of interest. This general rule is applicable because of the acceptance of the judgement of the rating agencies by market makers and by investors in municipal bonds. There was a time several years ago when municipal bond dealers would scarcely glance at the economic and statistical material presented by a municipal issuer but would submit a bid at public sale for bonds based entirely on the rating by one or more of the rating agencies. This utter dependence on the ratings seldom occurs today, not because of any lessening of the quality of the ratings, but because the SEC and various courts have held in recent years that the underwriter's responsibility to the investors to whom municipal bonds are marketed goes far beyond merely ascertaining the rating on the bonds. This is a good thing. For many years the rating agencies have been insisting that their ratings were an expression of opinion as to the investment quality of the bonds, not a recommendation to buy or sell such bonds. It is not good practice for the underwriter to place so much dependence on the rating that bonds are purchased or sold on that basis alone.

How much effect does the rating have on the interest rate demanded by the market? The following table shows an approximate range of interest rate differentials for various ratings, applicable to a bond due in ten years: (Standard & Poor's ratings)

Between	*Interest Rate Differential*[1]
AAA and AA	10 to 15
AA and A+	15 to 25
A+ and A	10 to 15
A and A−	10 to 15
A and BBB+	20 to 30
BBB+ and BBB	10 to 15
BBB and BBB−	15 to 25
BBB− and BB	50 to 75

[1] "basis points" or hundredths of one percent.

Variations occur in these general rate differentials due to the term of maturity of the bond issue, the size of the bond issue and the market acceptance of the issuer over past years. Bond maturities shorter than the 10-year example shown will exhibit a narrower rate differential, while longer maturities of 20 or 30 years will often show a wider spread of interest rates. Also, small bond issues under $2 million will often be priced and bear interest at higher rates to attract individual investors, whereas issues of $5 million to $30 million will be more competitively priced because of institutional investor interest.

In addition to interest rate variations due to differences in maturity and other factors, there frequently exists a modest variation in rates for a particular issue of bonds, as a result of simple differences of opinion between market makers or the investment bankers offering to purchase municipal bonds from the issuer. These differences of pricing opinions are the economic ingredients of competition, when a community receives sealed bids at the public sale of a bond issue. Frequently, there will be "spreads" between the lowest interest rate bid and the highest rate bid of up to 50 basis points, i.e. hundredths of one percent, for the same identical bond issue. These differences of opinion stem from, (1) differing ideas of the immediate future of interest rate trends, (2) from the individual bidder's knowledge of specific investors from whom he may already have a commitment to buy some or all of the bonds or, (3) in some instances, from the local, parochial interest of a community commercial bank wishing to acquire the bonds for its own investment portfolio.

INVESTOR USE OF BOND RATINGS

While most of the preceding discussion on the impact of ratings on parking bonds has been directed to the effect of ratings on the interest rate, as determined at the initial issuance of a bond issue, such bond ratings were originally conceived by the rating agencies and continue today to be designed for the use and assistance of the ultimate investor. The importance of these ratings to investors is amply shown in the provisions of the National Banking Act of 1933, which requires national banks to restrict their municipal investor portfolios to bonds rated not less than Baa by Moody's and BBB by Standard & Poor's, or if not rated, they must be able to justify an equivalent investment quality to the bank examiners. Thus, for commercial bank municipal investments, the rating accorded a proposed bond purchase for their portfolio, is of prime importance.

Other institutional investors, such as insurance companies, savings banks, and business corporations, do not have the same legal restrictions on municipal bond investments as the commercial banks. Generally, these other institutions set forth their own investment policies, most of which are based on the rating assigned to a municipal bond or the equivalent in quality as determined by their own analysts.

No such requirements exist for individuals buying municipal bonds. However, conservative, non-speculative individuals frequently restrict their purchasers to bonds in the higher rating categories, whereas high-flyers, looking for the highest possible tax-exempt yields, will buy non-rated, marginally secured bonds. In recent years, because of adverse publicity concerning some tax exempt bond issues, many individual investors have "upgraded" their investment policies in municipal bond purchases.

Changes in Ratings

The rating assigned by Moody's Investors Service and Standard & Poor's Corp. at the initial issuance of a particular bond issue or class of similar bond issues does not mean that the rating will remain unchanged during the life of those bonds. There are two basic premises for a change to occur in that original rating. The first and most common change is based upon the transition of a new project financed by bonds, the revenues from which are used to secure bonds, from the construction stage to completion and operation. Both rating agencies utilize revenue projections made by recognized engineers in partially determining the investment grade of revenue bonds. If the detailed information is adequate for a rating to be assigned to a new issue of revenue bonds to finance a new project, Moody's will preface their rating with a "P" symbol for provisional and Standard & Poor's will preface their rating with "CON" for construction project. These additional symbols will remain on the rating until such time as the project financed has been completed *and* a minimum of one year of revenue earnings history has accumulated. When a sufficient trend of such revenue earnings has been demonstrated, the prefacing symbol will be removed and, if the earnings are reasonably close to those projections made initially by the engineers, the rating itself will be continued. Obviously, if such earnings history is significantly different from that projected at the time of issuance of the bonds, the rating will be subject to a change.

The second and more significant premise for a change in rating is based on material changes in the investment quality of the bonds, as a result of a demonstrated decline in the earnings potential constituting the security of a revenue bond or in the tax base, economics, tax collections, and other factors affecting the ability of a municipality to meet payments on its general obligation debt.

With respect to parking system revenue bonds, the primary cause for a reduction in an outstanding bond rating is a continuing decline in both traffic and revenues of the pledged facilities. As mentioned before, the addition of on-street meter revenues will ameliorate any sudden decline in off-street parking revenues, but a declining trend in meter revenues also will be taken into account by the rating analyst. An increase in an existing parking revenue bond rating can usually only be attained, (1) by a demonstrated and continuing increase in traffic and revenues, (2) by the addition of debt-free, revenue producing parking facilities (with revenues pledged under the bond issue), or (3) by the steady reduction of revenue bond debt vs. a continuing or increasing revenue stream.

If parking bonds are general obligations of the municipality, the rating accorded such bonds will be virtually independent of the revenue producing capabilities of the parking project for which the bonds were issued. Thus, the financial record of the municipality as a whole will constitute the criteria for any change in the rating assigned such bonds. Increasing debt, budget deficits, stagnant tax base, or declining collections of property taxes are all reasons for a decline in a general obligation bond rating. Similarly, debt reduction, balanced or surplus budgets, a developing tax base and well enforced tax collections should give rise to an upward revision of a community's bond rating.

Changes in bond ratings, after the initial issuance of the bonds, have an immediate effect on the market price of those bonds in what is called the "secondary market." This over-the-counter market is made by banks and dealers who buy, sell, and trade municipal bonds sold into the market or purchased from the market by investors. The maintenance of a good secondary market is only important to the issuing community in the event that the community is planning a new issue of bonds or has a continuing program of purchasing specific issues of its bonds from time to time for sinking fund purposes. For smaller municipalities with serial bonds outstanding, which enter the new issue market with a bond sale every ten years, the secondary market in its bonds is of little importance. However, for larger cities and counties which are "in the market" at least once a year, the maintenance of a good rating and a strong secondary market for its bonds is essential.

SUMMARY

In summary, there are many practical reasons for every municipality or authority, planning the issuance of bonds to finance new parking facilities, to supply at least the two major rating agencies with complete and extensive information on the new financing as early as practicable. The rating agencies require a formal application to commence the analysis for rating purposes, and their fee is commensurate with the time and effort expended. All of the rating agencies have a policy of discussing their rating conclusions prior to the publication thereof. If the issuer disagrees with the rating, the agencies are available for a hearing concerning the communities' arguments and to examine any additional information made available.

There is an inverse relationship between bond ratings and new issue interest rates, i.e., the higher the rating the lower the rate of interest. Thus, it behooves any municipal or authority issuer of parking bonds to place behind those bonds the highest degree of security legally and economically available. Investors place a great deal of faith in the bond ratings of Moody's and Standard & Poor's particularly. The "name of the game" in the marketing of new parking financing is to create the highest degree of investor interest in the new bonds prior to *and* after their sale and the most effective way to accomplish this end is to obtain the highest rating possible.

CHAPTER 10

PARKING FEASIBILITY STUDIES

The need for a parking project feasibility determination should not be underrated. Its obvious purpose is to determine the projected functional and financial feasibility of the proposed development. Such a determination provides the basic evidence with which to attract investor interest and public support and on which final development decisions can be made.

SITE ANALYSIS

A common starting point for project development is identifying potential sites for the proposed parking project and defining the scope of the project. Many new parking facilities are developed as an on-site auxiliary to a specific host traffic generator, thus site selection for such a parking facility is influenced more by the locational needs of the generator. The locational relationship of the parking facility is especially dependent on the walking distance to the traffic generators it is intended to serve, as well as the site's vehicular accessibility. The analysis of alternate sites is necessarily interwoven with functional and aesthetic design considerations. Some preliminary design analysis is performed to properly assess the suitability of alternate sites. This might involve the potential placement of the parking structure on the site, examining the capacity and traffic control constraints of feeder roadways and street intersections serving the site, and identifying on-site vehicular and pedestrian circulation opportunities and constraints.

Site analysis also considers land availability and potential physical constraints other than access. Site cost, public policy and regulations, zoning and/or deed restrictions, property easements, and the character of surrounding development are among site availability considerations. Existing building demolition or relocation, drainage and utilities, excavation needs, relative ground and surrounding building elevations, soil foundation characteristics, and space for construction staging and the difficulty of handling traffic during construction are among the physical development constraints to be considered.

During this early stage of alternate site analysis, some preliminary economic feasibility determinations may also be made. Using national or regional averages, feasibility analysts will seek to quantify parking demand and needs, estimated costs, and expected financial performance with a rough pro forma analysis. This early pro forma analysis enables the project and potential alternate sites to be assessed and compared.

PROJECT FEASIBILITY

The requirements and expectations of new or expanded parking developments are influenced by market conditions anticipated for the host traffic generators expected to influence the proposed facility. Marketing data useful to the analysis of demand and need for a specific proposed parking project include:

1. the number of estimated person trips generated, including arrival and departure time patterns,
2. socioeconomic characteristics of potential patronage of the host traffic generators,
3. the stability and growth of the potential traffic generators,
4. the traffic generator's needs with respect to patron accessibility, passenger and parcel loading areas, access for delivery, service and maintenance vehicles, special security and/or convenience requirements, and regular vs. infrequent events affecting trip generation or parking space usage,
5. characteristics of other parking in the market influence area, especially rate structures and their ability to attract parking from the proposed facility,
6. expansion and development of adjacent areas,
7. availability and usage characteristics of adjacent public transportation alternatives.

This marketing information is used to develop parking demand forecasts and determine parking space needs to establish the optimum size of the proposed parking facility.

Utilizing preliminary construction cost estimates and the projections of demand for the proposed parking project, at parking rates consistent with the area, the analyst can project the available revenue stream anticipated to cover the expenses of operation and maintenance and repayment of indebtedness. At this point, potential funding sources can be identified.

PRIVATE DEVELOPER FINANCING

If the proposed parking project is part of a new traffic generator development, potential tenants will influence the project's financial feasibility. To address the concern, the developer usually establishes requirements for the type of tenants that will be sought. For example, projects containing significant amounts of retail space may seek one or more major "anchor" tenants, who will contribute toward amortizing the permanent loan for a project. For large office projects, the developer may seek one or more major tenants and prelease a large portion of the space before financing. Developments with less speculative predevelopment beginnings may already have an operating and/or management agreement or lease from a major tenant upon which the project was conceived.

Regardless of the method of financing chosen by a private developer, the existence of a well prepared economic feasibility study for a proposed parking project is a necessity for seeking financing. In the process of obtaining financing, a private developer may seek equity investors, perhaps through a limited partnership investment group or through the major tenants in the overall development project. In addition, any mortgage lender will require a complete study of the projected economics. The projected cash flow of the development, whether a single parking facility or an overall development of which the parking is an integral part, will be affected significantly by the analysis show in the parking feasibility study.

PUBLIC PARKING FINANCING

In any type of public revenue bond financing for parking projects, the existence of a well prepared economic feasibility study is necessary for the successful sale of such bonds. This necessity could range from such a study being used to prove the economic viability of a system parking revenue bond issue and thereby assuring an investment grade credit rating, to a legal necessity for the issuance of additional parity revenue bonds pursuant to an earnings test in the authorizing bond resolution or trust indenture for outstanding parking revenue bonds.

For public projects having benefits not directly measurable in dollars, a cost/benefit analysis, as part of the economic feasibility study, is a technique used to evaluate project worth above costs. Certain public-involved parking projects may require such a cost/benefit analysis to identify direct benefits to the community or the public, in exchange for the public involvement or any public subsidy involved. Such a need also should be demonstrated when a municipality issues general obligation bonds for a new parking facility.

If a new parking project is to be financed by a municipality with tax exempt bonds payable wholly or partially from tax receipts, as general obligation bonds, special assessment bonds or bonds payable from sales tax or hotel/motel tax receipts, the governing body of the municipality should be informed of the estimated amount of tax burden such a financing might impose on such tax receipts. It should know whether the proposed parking facility might generate enough revenues, by itself, to cover its own operating and maintenance expenses. The governing body will also want to know if the entire burden of annual debt service on the bond issue will have to be paid from tax receipts or whether all or some portion of such debt service can be met from the net revenues of other parking facilities owned and operated by the community.

Not only does this type of analysis constitute good municipal budgeting practice, but in some states it could have an effect on the legal debt incurring powers of the municipality. In New Jersey, for example, the Local Bond Law provides that a municipality may deduct from its outstanding tax supported debt, to the proportionate degree that user revenues support annual debt charges, all or such proportionate amount of such user supported debt from the calculation of the community's debt incurring margin for new debt issuance. In addition, the rating agencies approach their analysis of a municipality's outstanding debt in the same fashion, deducting from their calculation of net tax supported debt, all or a prportionate amount of those issues supported by user revenues.

THE FEASIBILITY REPORT

The sale and marketing of a tax exempt bond issue to finance a new parking facility requires the publication of an offering prospectus or "official statement." This official statement is a legal document, prepared by or on behalf of the issuer, which describes the bond issue completely, including the security behind the bond issue, the legality

14. determine and discuss the community's and the private sector's plans for economic development over the foreseeable future and the impact on parking demand in the area of influence of the proposed parking facility,
15. discuss and explain the assumptions and methodology used to project the parking demand at the site of the proposed parking project,
16. show the projected parking demand annually for each of the succeeding ten years, for the proposed parking facility, broken down between monthly permit usage, short term transient and all-day parkers,
17. justify the need for the proposed parking project, based on the projected parking demand on the proposed site, the recommended capacity of the new facility and the anticipated mix of parker utilization,
18. show a projection of parking traffic, gross parking revenues and average fee at projected rates, broken down by each publicly owned off-street parking facility, including the proposed new parking facility, and a projection of on-street meter revenues, at projected rates, over each of the succeeding ten years,
19. show a projection of operating and maintenance expenses, including any taxes and management fees, broken down by each publicly owned off-street parking facility, including the proposed new parking facility, over each of the succeeding ten years, indicating any assumptions of inflation factors,
20. show a summary projection of gross parking revenues from all sources, operating expenses and net revenues available for debt service, for each of the succeeding ten years, based on the detailed projections developed,
21. compare projected annual net revenues from all facilities to existing annual debt service on outstanding parking bonds, if any, plus estimated annual debt service on the proposed new facility financing, footnoted to show the rate of interest estimated, the amount of the new financing and the source of the estimates,
22. include, wherever applicable, charts as well as statistical tables of relevant data used,
23. include a detailed breakdown of the projected capital cost of the proposed new parking facility, obtained from the client community or the construction engineers.

The requirements to be included in a feasibility study for publication in an official statement, as previously shown, assume that the client municipality has existing off-street parking facilities and on-street meters combined in an operating parking system. To the degree that the proposed financing only includes a single, new off-street parking garage, not supported by other parking revenues, the same type of informational presentations would be required, only confined to the proposed new parking facility.

CHAPTER 11

ESTIMATING CONSTRUCTION DISBURSEMENTS

Private or municipal owners of planned new parking facilities are aware of the time that it takes from conception to completion of such projects. The planning stages for a new facility might take several years and involve the expenditure of sizable up-front amounts for surveys, studies, and plans. The actual, physical construction of a multilevel parking garage, however, ranges in time from about six months to over three years, depending on site constraints, the size of the garage, and the complexity of the design.

The disbursement of total project costs follows an established pattern, based on the realities of accumulated pre-construction development costs, land acquisition, disbursements for financing, and finally construction costs. By the time the designated contractor is given approval to begin construction, these preliminary costs have been paid, or the money set aside therefor, and the owner has left either a construction fund to draw down or a net balance of an interim loan to draw upon. Unless there has been included in the negotiations a guaranteed maximum price, a contingency fund for overruns is also included in these available construction moneys. Some contingency should be included to cover owner related work order changes.

At this point, the general contractor will submit to the owner a budget or estimate of his anticipated monthly draw down for the construction, i.e., progress payments. The contractor establishes this schedule of monthly payments based on his construction schedule of work to be performed on specific dates by all of the various subcontractors. The aggregate monthly invoice to the owner is based on the various line items from the subcontractors.

As the work proceeds, the general contractor submits a monthly invoice to the owner's representative for review. The owner's engineer will determine if the work described in the invoice has or has not been accomplished and will so certify to the owner, making a recommendation with respect to payment. At that point, the owner draws the money from an escrowed construction fund, held by a bond trustee, or draws upon his line of credit pursuant to an interim financing agreement with a commercial bank.

PROJECT DELIVERY METHODS

Within the parking construction industry, as with the construction industry as a whole, there are four basic methods for the construction process of delivering a parking project for the owner. These methods are:

1. Conventional bidding
2. Negotiated contract
3. Design/construct (design/build) and
4. Construction management

The value of determining the preliminary construction disbursements over the estimated elapsed time to completion will vary somewhat with the project delivery method utilized by the owner.

In the case of a turnkey procedure, such a preliminary evaluation would have little significance, because the design/construct contractor has already included the cost of this interim construction financing in his turnkey price. In the case of a multi-building complex, where the owner utilizes the construction management procedures, some value could be realized by an evaluation of the cost of money during the construction of the parking facility portion, but the degree of value would depend on the relative size of the parking facility in comparison to the total complex. In all other methods of project deliveries, the cost of money during construction is of significance to the owner, and the preliminary evaluation of the cost should be of value.

CONSTRUCTION DRAW-DOWN SCHEDULE

In 1984, the National Parking Association published a study entitled "Parking Facility Construction Payment Schedule Study." One purpose of this study was to ascertain if there existed any differences in the pattern of construction money paid out periodically during the construction of different parking projects. A conscious effort was made in the collection of data on 40 different parking projects, to determine the size, type of structural system, location and other details concerning the project. It was felt that three major factors might have a bearing on the flow of construction money: (1) the type of structural system, i.e., cast-in-place, precast or steel; (2) the size of the parking project, i.e., capacity of vehicles; and (3) the location of the project, i.e., northern vs. southern areas.

Because the total construction time of the 40 parking projects analyzed varied from 7 to 43 months, the data was reduced to a common time factor based on the percentage of the total time of construction. Increments of 5% were chosen as sufficient to establish a trend line comparable to the traditional monthly payment schedule.

The conclusions of this study indicate that a typical construction draw down curve for a parking project is a modified *S* curve. It showed a significant dispersion of the draw down amounts between the three types of structural systems analyzed, much less dispersion because of the location of the project and an insignificant amount of dispersion due to the size of the parking project.

The primary conclusion, however, showed that garage owners, both public and private, can estimate more accurately, prior to construction, the net costs of financing and the costs of interim financing, using the developed percentage curves rather than depending on a straight line monthly draw down.

Table 11-1 shows the average draw down for construction of all the 40 garage projects analyzed in the study, showing, in 5% increments of time, the percent of money paid out in each incremental period, the accumulation of money paid out and the converse of funds which would be remaining in a fully funded construction fund.

To illustrate the use of these construction draw down curves, assume a new parking garage project is to be constructed at a total hard cost of $25,000,000, not including land costs or soft costs of financing. Assume also, for ease of calculation, that the total construction time was 20 months. Using Table 11-1 (the average of all types of construction, location and size), a reasonably accurate estimate of the costs of interim financing for a private owner or the earnings from a funded construction fund for a public owner, can be determined. Assuming that a private owner would have to pay a commercial bank ½ of 1% as an interim loan commitment charge on the unused portion of a $25,000,000 line of credit for construction and 10½% on the accumulated funds drawn down, the total cost of the interim financing over the 20 months of construction, using the draw down schedule curve, would be $2,645,105, of which $86,492 would

Table 11-1. Typical Parking Garage Construction Draw Down Schedule[1]

PERCENT OF TIME	PERCENT OF MONEY PAID OUT	CUMULATIVE	
		MONEY PAID OUT	FUNDS REMAINING
5	3.6830	3.6830	96.3170
10	4.4865	8.1695	91.8305
15	4.8431	13.0126	86.9874
20	4.9618	17.9744	82.0256
25	5.5389	23.5133	76.4867
30	8.5099	32.0232	67.9768
35	6.8310	38.8542	61.1458
40	7.9299	46.7841	53.2159
45	7.6557	54.4398	45.5602
50	6.7842	61.2240	38.7760
55	6.4271	67.6511	32.3489
60	6.3394	73.9905	26.0095
65	4.4181	78.4086	21.5914
70	5.1424	83.5510	16.4490
75	4.1460	87.6970	12.3030
80	3.2752	90.9722	9.0278
85	2.4366	93.4088	6.5912
90	2.7102	96.1190	3.8810
95	2.0557	98.1747	1.8253
100	1.8253	100.0000	—0—

[1] Average of 40 projects analyzed.

consist of the loan commitment charge. Similarly, if a public municipal owner funded a construction fund of $25,000,000 and invested the total amount at an average of 8.10% in U.S. Treasury obligations, the aggregate earnings, using the draw down schedule curve, would accumulate to $1,401,214, which could be used to offset the interest paid on bond financing during the 20-month period of construction.

If either a private owner or a municipality used the traditional straight line, equal monthly draw down of construction costs, the projected cost of interim financing for a private owner would be underestimated by about 15% in the aggregate and the municipality would overestimate the investment earnings from the construction fund by a similar amount.

CHAPTER 12

SIZING A MUNICIPAL PARKING BOND ISSUE

Calculating the amount of a proposed tax exempt parking bond issue can be done on a calculator; for the final bond offering, however, it should be done with the help of commercial bond sizing programs for computers, which most banks and investment banking firms have available.

ASSUMPTIONS

Certain assumptions or factual data are required to do the calculations. These are as follows:

1. *Aggregate Construction Costs*—the estimated or contractual amount of the hard cost of completing the parking project, including architectural and engineering costs, feasibility studies, permitting fees, land costs, site preparation, filing fees, supervision during construction, and construction contingencies.
2. *Costs of Issuance*—the estimated or contractual amount to be charged to the issuer for bond counsel, local counsel, accountants, printing of the preliminary and final official statement, rating agency fees, trustee fees, and closing costs.
3. *Time of Construction*—the estimate of the number of months from the date of bond issue closing to the completion of the facility (certificate of occupancy), including either a draw down schedule in dollars from the contractor or reliance on the draw down curve described in Chapter 11 (see Table 11-1).
4. *Capitalized Interest Period*—a determination of the number of months from the date of the bond issue for which interest on the bonds will be capitalized or funded from bond proceeds; usually not less than the construction period.
5. *Bond Interest Rate*—while the input to a computer sizing program will accept multiple interest rates for serial bonds or a combination of serial maturities and term bonds, such calcula-

tions become overwhelmingly burdensome when calculating a projected bond size with a hand calculator; when calculating by hand, use an average rate of bond interest; for most serial maturity, level debt service bond issues a rate of interest about 10 basis points (1/10th of 1%) lower than the rate assigned to the longest maturity of the bond offering will suffice as an approximate average interest rate.

6. *Underwriter's Spread*—the estimated or contractual amount, usually expressed as a percentage of the par value of the bond issue, representing the discount from the par value of the issue at which the underwriter agrees to purchase the bonds (varies significantly depending on the size and credit rating of the issue).
7. *Debt Service Reserve*—usually required for all types of revenue bonds, this represents the amount of bond proceeds set aside into a debt service reserve fund, to be retained and maintained over the life of the bond issue, to provide limited bondholder protection against a possible default from insufficient revenues being available to pay current debt service requirements; a typical debt service reserve fund is capitalized or funded from bond proceeds in an amount equal to either the maximum or average annual principal and interest requirements of the bond issue over the life of the bonds; the mathematical formula for calculating the average annual principal and interest requirement is $1/a_{\overline{n}|i}$, where a is an annuity due over period n at rate i, which results in the periodic payment necessary to pay off a loan (this is the reciprocal of the present worth of 1 per period).
8. *Bond Insurance Premium*—if bond insurance is purchased to enhance the credit of the bond issue, the amount of the premium charged by the insurance company is a one-time charge, which usually is payable from bond proceeds; the estimated or quoted rate from the insurance company is a percentage of the aggregate amount of principal (initial par value of the bond issue) and interest on the bonds over the complete life of the issue, less the amount of capitalized interest set aside from bond proceeds to pay interest during the construction period; the quoted percentage rate will vary from ½ of 1% to over 1% depending on the basic security of the bond issue.
9. *Investment Rate*—due to the arbitrage restrictions in the 1986 Tax Law, as recently promulgated in Treasury Department regulations, bond counsel have advised issuers to calculate invest-

ment earnings from tax exempt bond proceeds at a rate not exceeding the average rate at which the bonds are sold; for purposes of calculating an estimated bond size, it is prudent to use the bond interest rate as previously described.

CALCULATIONS

The formula for calculating bond size consists of solving an algebraic formula with only one unknown, i.e., the size or par value of the bond issue or P. To achieve this end result is merely the addition or subtraction of the various input data and assumptions shown above.

To illustrate the basic bond sizing formula, assume the following:

1. C = aggregate construction costs,
2. CI = the total costs of issuance,
3. CT = the time of construction in months from the date of closing of the bond issue,
4. CIP = the capitalized interest period, in months from the date of the bond issue,
5. R = the average interest rate to be borne by the bond issue,
6. U = the percentage of underwriter's spread,
7. BP = the percentage of premium to be charged for bond insurance, if any,
8. IR = the investment rate to be used for the investment of funds (for tax exempt financing, this should be no greater than R).

First, an independent calculation should be done to determine the earnings from the investment of the construction fund moneys. Utilizing the draw down schedule methodology described under "Estimating Construction Disbursements" in Chapter 11, with the construction cost C, at the investment rate IR, the estimated amount of investment earnings from this source can be determined.

Assume the resulting amount is CE, or construction fund earnings.

The calculation of the amount to be set aside from bond proceeds to pay interest on the bonds during at least the construction period of the parking project can be expressed by the following formula:

$$R/12 \times CIP \times P = \text{the amount of capitalized interest or A}$$

The calculation of the amount of investment earnings to be achieved from A is somewhat more difficult. Once a bond issue has been sold and delivered, the interest payments are constantly accruing and are paid out semiannually by the bank paying agent to the bondholders from the moneys in the capitalized interest fund, so long as funds remain therein for the purpose.

The second independent calculation which should be made is to determine that portion of the overall bond sizing formula which represents the investment earnings from A or the capitalized interest fund. For purposes of illustration, the 20-month construction period and the average interest rate of 8.10% used in the example shown in "Estimating Construction Disbursements" is used here to develop the investment earnings, assuming that the CIP or capitalized interest period is also 20 months. Adjustment of the final segment of the formula can be made if the CIP is longer than the period of construction. All of the letter designations have been defined previously, with P representing the par value of the proposed bond issue and A representing the total amount of capitalized interest.

Time Period	*Investment Rate*	*Times*	*Money to be Invested*
1st six months	IR/2	×	A
2nd six months	IR/2	×	A − (R/2 × P)
3rd six months	IR/2	×	A − (R × P)
remaining months	IR/12 × (CIP − 18)	×	A − (1.5 × R × P)

To reduce the addition of these four elements to a workable, single element to be included in the bond sizing formula, it is easier to assign the known values of the illustration being used; i.e., IR = 8.10%, R = 8.10% and CIP = 20 months. Thus, the four periods would look like the following:

Time Period	*Investment of Moneys*
1st six months	0.0405 × A
2nd six months	0.0405 × (A − (0.0405 × P))
3rd six months	0.0405 × (A − (0.0810 × P))
remaining months	0.0135 × (A − (0.1215 × P))

In the final bond sizing formula, all unknowns must be reduced to a relationship of a single unknown, in this case P or the par value of the bond issue. In the determination of A, the formula R/12 × CIP × P was used. By using the above mentioned illustrative values for R and CIP, the value of A can be expressed as 0.1350 P. By substituting this value for A in the previous table and combining algebraically, the following can be ascertained:

Time Period	*Investment of Moneys*
1st six months	0.0055 P
2nd six months	0.0038 P
3rd six months	0.0022 P
remaining months	0.0002 P
TOTAL	0.0117 P

Thus, given the illustrative bond interest rate of 8.10%, the investment rate of 8.10% and the capitalized interest period of 20 months, the amount of investment earnings from capitalized interest funds (AE) can be represented in the bond sizing final formula as 0.0117 P or 1.17% of the proposed par value.

The portion of the bond sizing formula which represents the amount of the debt service reserve (DSR) to be included is described above as $1/a_{\overline{n}|i}$, where a is an annuity due over period n at rate i. This is a figure which can be looked up in any book of annuity tables or calculated on a financial calculator. In the illustration being used here, the rate i in the annuity table would be 8.10% or R as used in this bond sizing formula.

One assumption, however, has not yet been made. The amortization period of the proposed bond issue must now be assumed. Most parking bonds begin the first principal retirement or sinking fund allocation at least one year following the estimated completion of the parking project. This policy is followed to allow a year of operation and the generation of parking revenues to occur before money has to be set aside for principal amortization. Thus, assuming that the longest or final bond maturity would be 30 years from the date of the bond issue, the first payment of principal or sinking fund should occur about three years from the date of the bonds, leaving 27 years of annual amortization.

Thus, going into the annuity tables with 27 periods of amortization or n and a rate of 8.10% or i, the amount of the debt service reserve to be included in the bond sizing formula can be shown as 0.0923 P or 9.23% of the proposed par value of the bond issue. The investment earnings (DSRE) of this debt service reserve fund (DSR) during the 20-month construction period can be represented as follows:

$$0.0810/12 \times 20 \times .0923 \text{ P or } 0.0125 \text{ P}$$

where P is the proposed par value of the bond issue.

At this point, the following three additional assumptions must be made in order to complete the bond sizing formula:

1. the percentage of underwriter's spread to be charged (U); this will constitute a discount from the par value of the bonds; in this example a percentage of 1.5% will be used,
2. an amount of money estimated to cover the various costs of issuance of the bonds, as previously described (CI); in this example a figure of $100,000 will be used,
3. assuming that a policy of municipal bond insurance will be purchased by the issuer, the one-time premium cost can be paid from the bond proceeds; in this example a percentage premium of 0.75% will be used (BP).

Thus, the portion of the bond sizing formula which represents the underwriter's spread or discount can be shown as 0.015 P, where P is the par value of the bond issue.

Similarly, the bond insurance premium portion of the bond sizing formula is represented as follows:

$$0.0075\ (P + (R \times P \times 30) - 0.1350\ P)$$

Where 0.0075 is the assumed insurance premium rate (BP), P is the par value of the proposed bond issue, R is the 8.10% average rate being assumed for the bonds, 30 is the entire life of the bond issue in years to maturity and 0.1350 P is the previously calculated amount of capitalized interest (A) to be deducted. Assigning the illustrative values shown and combining algebraically, the resulting portion of the bond sizing formula representing this bond insurance premium (BI) can be shown as 0.0247 P.

Finally, all of the factors previously discussed can be brought together in the following simplified formula:

$$C + CI + A + U + BI + DSR - CE - DSRE - AE = P$$

Each item must be reduced to a relationship of P or an amount in dollars, in order to solve for P.

Using the same amount of total *hard* construction costs of $25,000,000, as shown in "Estimating Construction Disbursements" in Chapter 11, the final bond sizing formula can now be assembled, as follows:

Function	*Amount or % of P*
Additions	
Construction costs	$25,000,000
Costs of issuance	100,000
Capitalized interest	0.1350 P
Underwriter's spread	0.015 P
Bond insurance	0.0247 P
Debt service reserve	0.0923 P
Subtractions:	
Construction fund earnings	$1,401,214[1]
Debt service reserve earnings	0.0125 P
Capitalized interest earnings	0.0117 P

[1]Refer to "Estimating Construction Disbursements" Chapter 11.

Combining algebraically, the following formula results:

$$\$23{,}598{,}786 + 0.2428P = P$$

Solving for P shows that the size of the proposed bond issue (par value) should be exactly $31,297,921, which, when rounded to the nearest $5,000 bond denomination, means a bond issue amounting to $31,300,000.

To prove the calculations, substitute the bond issue size of $31,300,000 for P in the schedule above, as follows:

Function	*Amount*
Additions	
Construction costs	$25,000,000
Costs of issuance	100,000
Capitalized interest	4,225,500
Underwriter's spread	469,500
Bond insurance	773,110
Debt service reserve	2,888,990
Subtractions:	
Construction fund earnings	$1,401,214
Debt service reserve earnings	391,250
Capitalized interest earnings	366,210
Add balancing amount for rounding	1,574
TOTAL	$31,300,000

It should be noted that no investment earnings have been attributed to the amounts included for costs of issuance, underwriter's spread, or the bond insurance premium, inasmuch as it is assumed that these moneys will be paid out immediately following the closing and delivery of the bond issue.

For purposes of determining financial feasibility, in conjunction with pro forma income projections shown in an economic feasibility study, the amount calculated for the debt service reserve fund, as deposited from bond proceeds, can be utilized as the average annual debt service or average annual principal and interest requirements for the bond issue.

Also, it will be noted that no attempt has been made in this simplified calculation to take into consideration the interest on interest earnings. Commercial computer bond sizing programs do take this into consideration.

The same concept of calculations, as shown here for a tax exempt municipal bond issue, can be utilized for conventional mortgage financing, provided allowance is made for the cost of interim financing instead of capitalized interest, the exclusion of a debt service reserve and the exclusion of investment earnings. Based on the percentage of debt to equity and the estimated return on equity required, a total annual cost can be calculated to ascertain financial feasibility. Most large investment banking firms have developed computer models which will provide an accurate sensitivity analysis for financing under a variety of rate and cost conditions.

CHAPTER 13

THE ROLE OF THE PARKING CONSULTANT

The value of an economic feasibility study prepared by a qualified firm of parking consultants was shown in Chapter 10, as it applies to the determination of viable parking financing. However, the successful completion of a new parking facility and ultimately the successful retirement of its financing as planned, also depend to a significant degree on the planning and design talents of the engineers in the conceptual stages. The functional and detailed design of parking facilities has, over the past 30 years, become a highly specialized technical art, which is understood and practised by a relatively small group of parking consultants throughout the country.

While many parking administrators or corporate service officers may be able to determine, with sometimes uncertain accuracy, the need for local parking requirements and while a local architect may be able to design a structure in which to park cars, neither has the broad and detailed experience to plan for and design the most efficient parking facility to suit the client's needs and pocketbook. By virtue of his/her research, training, experience, and opportunity, the outside parking consultant has become an essential resource of the entire parking industry.

In these days of increased SEC vigilance concerning the disclosure of material facts about any new public or corporate financing, it is incumbent upon underwriters to investigate beyond the obvious forms of security and look into the use to which the funds to be raised are to be spent. Lenders also are much more aware of how their money is to be spent. In the case of new parking facilities, both underwriters and lenders will accept the technical expertise of a qualified parking consultant who can demonstrate that the proposed project is well planned and designed to perform as anticipated.

In this respect, both lenders and underwriters, as well as the rating agencies, are well aware of the fact that not all "consultants" are qualified as parking consultants. Because of the growing demand for new parking facilities throughout the country, so-called consultants, frequently one-man shops, are proliferating, often with little or no experience in the parking field. First, it is important for a client to recognize the difference between a qualified independent consultant and the knowledgeable salesman who may represent a particular parking product or system. The independent parking consultant will not benefit financially from the client's purchase of any parking product; on the contrary, any recommended specifications will eliminate those products or systems not applicable to the client's project, while permitting those that do apply to compete for the business.

Secondly, it is important for a client to recognize that even within the fraternity of able, qualified parking consultants or consulting engineering firms, there exists a diversity of functional specialties, within the overall range of parking skills needed for any one parking project. Some parking consulting firms possess a wide range of skills and experience in the parking field, while others specialize in a particular area of expertise. As an example, many highly qualified firms perform demand and economic feasibility studies but may not be qualified to do the engineering design work for a new garage. Others have vast experience in the design field but are not prepared to advise the client concerning the operations of a parking system.

No one parking consulting firm is qualified in *all* of the following functional areas in the parking industry:

- demand and feasibility studies
- functional design
- site evaluation
- detailed civil engineering
- revenue control evaluation
- management and operation
- restoration evaluation
- financial options

CLIENT NEEDS

The need for the services of a qualified parking consultant depends on the needs of the client in the above mentioned functional areas and the client's ability to provide such needs within the client's organization. Large municipal parking systems with adequate and experienced staff could perform some of the functional needs required in the planning and evaluation of a proposed new parking facility.

Large development companies also have the experienced staff to do the same planning and evaluation. In both instances, however, the detailed design of a new parking structure should still be left to a qualified parking engineering firm. When obtaining financing for such a project, lenders prefer an independent firm to design and supervise the construction of the project, in order to assure that their invested money is well spent.

The principal reasons for retaining one or more qualified parking consultants include the following:

1. to assist the client to focus on the development of a specific plan to solve a parking problem,
2. to supplement overburdened or inexperienced personel to accomplish the work to be done,
3. to make available to the client the best expertise in the area of parking required, and
4. to enable the client to present independent recommendations and conclusions, in order to assist in financing the project.

Frequently, municipalities can call upon existing personnel in the area of public works, traffic engineering and planning to provide excellent, basic data which can be used by a good parking consultant. Such assistance can reduce the time spent by the consultant in gathering needed field data, thereby reducing the fees and charges to the client. Given a recognition of the professional capabilities of the parking consultant and of the client's staff and personnel in their respective areas, a good working relationship is frequently developed that produces optimum results. In addition, because the parking consultant has a wide range of experience of observing success and failure over a wide range of parking problems and has data available from previous parking problems, he/she can focus on the client's needs and problems quickly, without the need for unnecessary or extensive basic research.

SELECTING A PARKING CONSULTANT

There are several phases in the process of defining, selecting, and working with a parking consultant, that should be borne in mind by a client. The first phase in this process is for the client to determine the broad concept of what they want the consultant to do. Is it merely to review an architect's design of a new parking structure; is it to perform a complete downtown parking and traffic demand study, with recommendations of sites and capacities for new parking projects; is it for the design and supervision of a parking project, for which the site, demand, and capacity are already established? Based on the client's personal knowledge, from recommendations from others or from a listing received from the National Parking Association or the Institutional and Municipal Parking Congress, the client should send a letter to those firms selected, requesting an expression of interest, based on as much specific information as may be available to the client at that time concerning the project. Such a letter should include a description of the work to be done; i.e., demand and feasibility study, review of existing design plans, new functional and final design work, or a combination. It should detail whatever preconceived plans the client has with respect to the project. It should ask for a description of previous work done by the consultant in the same functional area of work requested, and for details concerning the consulting firm, including the size of staff, biographies of personnel to be assigned to the client's project, and an indication of the availability of such personnel to begin the client's work on a specific date.

Unless the client knows exactly what work is to be performed by the consultant, it is best to indicate in this request of interest, the broad scope of what is expected. One method of eliciting more definitive responses would be for the client to request a suggested scope of services from the consultant, such as will accomplish what the client wants.

In this preselection process, the client should not attempt to assign a maximum dollar amount for the services of the consultant. It is better to request an approximate range of cost for the scope of services recommended by the consultant.

Once a reasonable number of responses of interest have been received by the client, each should be studied to determine how closely the consultant followed the instructions, what their scope of services includes, and whether that which is proposed will accomplish what the client wants. Only when firms of comparable qualifications and comparable scopes of services are evident, should the client compare the range of fees quoted. At this point, the client should schedule personal interviews, of not less than one hour each, with the top three or four consulting firms which have responded. The client should be prepared to ask questions relating to the firm's degree of interest and whether that firm was willing to formalize their interest in a contract for services.

It is most important for the client to schedule the personal interviews promptly after the receipt of the letters of interest but allowing those firms chosen adequate time for preparation. Also, it is important, following those interviews, to inform *all* of the firms making a presentation of the decision of the client as to the final selection. This should be done as promptly as possible.

THE WORK PHASE

Upon retaining a parking consultant, it is advisable for the client to assign a senior level person to act as the coordinator to represent the client. Particularly with large corporate or municipal organizations, a liaison between the consultant's staff personnel and the various persons of the client's organization who may be called upon to provide technical data is necessary. By frequent meetings between the assigned coordinator and the consultant, legal and political roadblocks can be ascertained and possible changes in the scope of the work can be defined. Such meetings will also serve to keep the client up-to-date on the progress of the work being done by the consultant. If the consultant's work is to be done in stages, each stage should be submitted to the client on time, through the assigned coordinator, in draft form for review by the client. This review should be done promptly, with recommendations made to the consultant so that if changes are required, the overall time schedule will not be affected.

Only when all of the client's staff recommendations have been incorporated or good reasons given by the consultant for not including them, should the final report be submitted to the client.

All qualified parking consulting firms value their reputation in their respective fields. Usually they will go out of their way to satisfy the reasonable demands of a client for whom they are providing a service. Their stock in trade is the recommendations they receive from past clients. However, the client should not expect a consulting firm to perform work clearly outside the agreed scope of services, without making arrangements for additional fee payments.

THE TEAM APPROACH

The services of parking consulting firms range from comprehensive CBD traffic and parking studies, with optional financing recommendations, to detailed engineering plans, working drawings and specifications and supervision of the construction of new parking structures and to management and operational structures for city-wide parking systems. Dependent upon the magnitude of the client's needs, more than one profession may be retained under the umbrella of the parking consultant. If required, separate engineers, architects, planners, landscape architects, lighting specialists, parking operators or others may be included in the professional consulting team to accomplish all that the client has requested. When such a team approach is used, the profession bearing the primary responsibility is usually the team leader and is accountable to the client.

There are no licensing or registration requirements for parking consultants performing demand and feasibility studies for either public or private clients. However, when the scope of services includes civil engineering and design, then the senior engineer in charge of the work must be registered in that state as a professional engineer. All states require that final plans, working drawings and specifications, used to obtain a building permit for a new parking structure, must be signed ("sealed") by a professional engineer or architect licensed in that state.

Most of the time, the nature of the client's work to be performed will determine the type of parking consultant to be employed. For a demand and economic feasibility study, it would not be necessary to include a firm with particular expertise in design work or parking management. If, however, the client had already accomplished that sort of preliminary investigation and was ready to proceed with the construction of a new parking structure, then a parking consultant with heavy experience and capabilities in design and supervision of construction and with resident professional engineers or architects registered in the client's state, would be required. It cannot be overemphasized that the client's best interests will be served by retaining a team of parking specialists of varied disciplines, dependent upon the complexity of the client's overall needs and the project to be accomplished. Not only will this approach produce the most efficient operating parking project, but it will give significant comfort to those investors financing the client's efforts.

The size of the parking consulting firm being considered by a client should not be a factor in the final selection process, contingent, however, upon the scope of work required. A small "boutique" firm, which has established a good reputation in one or two specialized areas of parking, can be as valuable to the client as a large firm of consultants. On the other hand, if the client's requirements include a wide-ranging scope of services from demand studies through design and into operational management, then the large consulting firm, perhaps with one or two specialty subcontractors, would be better serve the client's interests.

ENGINEER OR ARCHITECT

Should a client considering a new parking structure retain an engineer or an architect to design the building? It is the consensus of parking operators throughout the country, private and public, that parking facilities, particularly multi-level garages, designed by qualified parking consulting engineers are more successful from the standpoint of efficiency of operation and customer acceptance, than those designed solely by architects. There is a good reason for this. A parking structure should be designed from the inside out, beginning with the most efficient and customer-pleasing functional design, including the parking space layout and the interfloor circulatory system; this aspect is particularly important in those parking structures intended to serve coliseums, stadiums or convention centers where "dumping" or exiting time frames are critical. Frequently, architects are more concerned with developing a good looking, esthetically pleasing structure, with less regard to the internal functioning of the facility for parking automobiles.

By placing the leadership for the design of a new parking structure under the responsibility of a qualified parking consulting firm, a more cost efficient design will result. In most new parking projects, structural engineering, mechanical and electrical components, lighting, power, plumbing and heating comprise about 85 to 95% of the total cost of construction. The remaining 5 to 15% of the construction cost is for architectural work, primarily on the exterior facade of the building. It is apparent then, that a parking structure is primarily an engineering assignment, with architecture being the subsidiary profession. The client's requirements, with respect to the exterior treatment of the parking structure, can be accomplished by the inclusion of the client's architect or an independent architect acceptable to the client, on the professional team headed by the parking consulting firm.

SUMMARY

What has all this got to do with the financing of parking facilities? Regardless of the method of financing for a new parking facility, whether public or private, it is incumbent upon the owner looking for financing to create the most cost effective facility possible, in order to present to lenders the most attractive financial position possible. If the borrower is a developer or an industrial corporation, the responsibility of management to produce a well designed, operationally efficient parking project, ultimately is answerable to the partners or shareholders of the company. If the borrower is a municipality, elected officials and staff personnel are ultimately answerable to the taxpayers (and voters), whether the project is financed with revenue or general obligation bonds. The economic feasibility of a proposed new parking project is a function of demand, rates *and* the capital cost, as estimated by qualified parking consultants. This determination is the single most important factor to lenders concerned with the security of their investment.

CHAPTER 14

THE ROLE OF THE FINANCIAL PROFESSIONAL

The capital needs and economic risks associated with new parking investments have grown dramatically over the past years. Pressure from increasing economies of scale, technological changes, and inflation have risen to the point where many new parking investment undertakings are simply beyond the means of traditional, direct bank financing techniques. As a result, the need for a financial professional to assist both private and public owners of new parking facilities has become increasingly important.

The term "financial professional" includes a variety of professional functions or services provided by a firm or individual specializing in the development of funding for a client which will satisfy lender requirements while minimizing the cost and degree of effective recourse to the client. Dependent upon the type of client (public or private), the type of offering to be made (public bid or negotiated sale), and the relationship of the professional to the client (underwriting or advisory), the financial professional can assume any of the following roles:

1. financial advisor or consultant
2. investment banker or underwriter
3. merchant banker or owner-participant
4. mortgage broker
5. venture capital finder

These basic functions of the financial professional frequently overlap to some degree. While the investment banker and the merchant banker are basically performing their services to the client for their own account; i.e., acting as a principal in an arms-length transaction, they frequently offer advice to their clients in the process of developing a specific financing package for marketing. On the other hand, the financial advisor and the mortgage broker perform a fiduciary function and do not themselves act as a principal in the financing, for their own account. The venture capital finder often participates, along with the investors from whom he obtains commitments, to purchase the equity in the project financing package.

THE FINANCIAL ADVISOR

The financial advisor or financial consultant is a professional services employee of the client and performs his services in a fiduciary capacity, i.e., he performs for and on behalf of his client and no other. Both legally and ethically the financial advisor cannot and should not make any recommendations to his client, which would in any way benefit him personally. When recommending other firms to his client, such as engineers, accountants, law firms or investment bankers, the financial advisor should always make several recommendations, based solely on the reputation and expertise of the firms.

Particularly with respect to the proposed financing of new parking facilities, the financial advisor can offer the following services to public and private clients:

- the analysis of alternative methods of financing
- structuring debt financing
- structuring equity financing
- development of long term public debt programs
- analysis of bond indentures
- critique of economic feasibility reports
- preparation of offering prospectae
- development of capital leasing programs
- presentations to rating agencies
- presentations to bond insurance companies
- presentations to governing bodies at public hearings
- expert testimony at bond validation proceedings
- representing the client to underwriting firms
- working with attorneys on legal documentation
- analysis of market conditions for financing
- recommendation of qualified parking engineers
- recommendation of qualified bond attorneys
- scheduling and expediting new financing

The value of a financial advisor to his client lies in the knowledge and expertise of the financial advisor in the specific business of the client, i.e., parking, *and* in knowing the requirements of lenders.

Contingent upon the degree of sophistication of the client in the area of financing, the role of the financial advisor is often that of an educator. Convincing clients of the hard realities of the security required by lenders for parking financing can be the most difficult aspect of a financial advisor's work.

THE INVESTMENT BANKER

The relationship of an investment banking firm or underwriter to the client may be characterized as "arms length." It is the opposite relationship to that of the financial advisor. While the investment banker may offer the client advice during the process of developing a new parking project financing, it is gratuitous. The investment banker only earns his fee for services when the final financing package is completed and the required funds are made available to the client.

Project financing by an investment banking firm has been defined as:

> "The financing of a project in which the lenders are satisfied to look initially to the cash flow and earnings of that project, as the source of funds from which the loans will be repaid, and to the assets of the project as collateral for the loans."

The merit in this approach is that financing is arranged on the basis of the viability of the project itself, with consequent limited impact on the credit worthiness of the sponsoring company or public entity. In the area of financing parking projects, however, the economic viability of a single parking project is seldom sufficient to enable the financing to be secured by the project alone. Dependent upon the degree of risk which lenders are willing to assume, project financings are classified as full, limited or non-recourse financing. At one extreme, lenders assume no project risk; at the other extreme, lenders assume all of the risk, with no recourse to project sponsors. By tailoring a parking financing to the particular characteristics of a specific project, an investment banking firm can take full advantage of the variety of requirements and incentives offered by different lenders, thereby optimizing the overall terms of the financial package offered to their client.

Among the services provided by an investment banking firm to their parking clients, are the following:

- provide financial feasibility analyses, based on economic and demand studies available

- formulate financing plans, including the projected impact on the client's credit ratings
- develop alternative financing structures, including legal, tax and credit elements
- evaluate available capital market alternatives
- seek equity participants
- raise both equity and debt capital
- negotiate the best available terms for the client
- expedite credit rating negotiations

Most of the larger investment banking firms have developed customized computer models, which can be utilized for their clients project financings. The flexibility of such program software enables the investment banking firm to produce a fully operative project model. This enables the firm to produce sensitivity analyses for their client for a specific project and to easily change such variables as construction costs, debt/equity profile, interest rates and maturities, rates of return on equity and tax considerations. Using these models, the investment banking firm can analyze a client's project from the perspective of the lender, equity investor and the project owner, to determine the financial feasibility of the project.

Inherent in this role of the investment banker is the premise that all of the financing which may be developed for a parking client will be marketed by the investment banking firm to other investors, whether other equity participants or other purchasers of debt obligations. The investment banking firm, when acting as a underwriter, purchases all of the debt or equity from the client directly and markets such obligations to investors. The sole profit to the investment banking firm is the difference between the purchase price paid to the client and the sale price to investors.

THE MERCHANT BANKER

Investment banking firms, commercial banks and some savings and loan institutions sometimes act in the capacity of a "merchant banker." By definition, this means that the bank or banking firm is acting wholly or partially on its own behalf, as the ultimate investor. Such a merchant banking relationship with a client is even more at "arms length" than the role of the underwriter as described above.

Although the same financial services are provided by the merchant banker, the tendency is always toward a more secure investment for the benefit of the merchant banker as investor.

Frequently, and subject to the overall size of the financing required, the lead merchant banker will negotiate all of the terms and conditions of the financing package with the client and then "syndicate" a substantial portion of the financing with a group of similar institutions. The lead banker decides which other institutions to approach, negotiates the terms of participation with each institution individually and decides which institutions to include in the final purchase. Such a syndicated financing approach excludes the borrower from the key aspects of the transaction by turning over the writing of the terms of the financing and the negotiations with the "marketplace" to a lead merchant banker. Only those sophisticated client-owners who have the expertise to retain control of the ongoing negotiations should entertain such an arrangement.

While in most cases this traditional syndication approach leads to satisfactory financing for a new project, it rarely leads to the optimal financing in the best interests of the client. Primarily, the lead merchant bank faces an unavoidable conflict of interest in proposing the terms for a syndicated loan. As a lender, the merchant bank would like to have as much recourse to the project sponsors or owners as possible, and at the highest "spreads" obtainable. The merchant bank may propose terms which are sufficient to win the mandate from the client borrower, but not necessarily the best obtainable in the marketplace. Therefore, the lead merchant banker has a conflict of interest in attempting to advise the client on how to get the best terms available. It may be difficult for the client to differentiate between terms the lead merchant bank would like to have as a lender and terms that are really necessary to syndicate the transaction.

THE MORTGAGE BROKER

Mortgage brokers may include large investment banking firms, small local investment banking firms, mortgage companies, or individuals specializing in finding mortgage financing for clients.

Contingent upon the size of the firm retained as a mortgage broker, financial advisory services may or may not be available to the client borrower. Usually, the larger firms and mortgage companies will confer with the client and assist in developing a term sheet of the requested mortgage. In addition, such firms will prepare a formal memorandum describing the client, the project and its feasibility, and the anticipated terms of the mortgage obligation.

Again by definition, a mortgage broker does not assume any underwriting liability for the borrowing. It is a classic example of "best efforts" financing, i.e., if the mortgage broker can find a lender, at terms satisfactory to the client, then the transaction can be consummated.

It is traditional in the mortgage brokering business for an exclusive agreement to be executed between the client and the broker, similar to that requested by real estate brokers. Dependent upon on the size and reputation of the mortgage broker, this exclusive agreement can be beneficial to the client. Basically, such an agreement prevents the client's mortgage financing from being "shopped around" by a variety of brokers to a variety of potential investors at a variety of terms. The client borrower has a much greater degree of control over this transaction with such an exclusive arrangement. The primary client risk, however, is in the initial choice of the mortgage broker.

THE VENTURE CAPITAL FIRM

Privately owned parking projects are seldom financed with 100% debt. While mortgage financing may provide from 50 to 90% of the completion costs of a new project, the balance must be comprised of equity. Such equity contributions may consist of the land and development costs but frequently consists of cash contributions as well.

Major parking operating companies usually contribute such cash as may be required, over a mortgage commitment, from general corporate resources. Developers, however, are more prone to raise equity contributions from institutional investors or limited partnership groups.

It is in this latter area of raising equity that the venture capital firm can be of assistance to a client borrower. The venture capital firm offers its expertise in structuring an equity offering, writing the necessary offering memorandum, advising the client as to the anticipated return on equity demanded in the market and making the approach to equity investors. For a specific parking project financing, such an offering would probably be a limited partnership offering.

Not the least of the service offered by a venture capital firm lies in its knowledge of particular equity investors or groups of investors, who have an "appetite" for the type of project proposed by the client. Equity investors, whether institutional or individual, have generic likes and dislikes, with respect to the basic industry in which they invest their money. As mentioned previously, the parking industry is not looked upon as a "glamour" industry, therefore equity investors are not beating down the doors of parking companies or developers.

Only a very few venture capital firms will be equity participants in new parking projects.

Of primary concern to a parking client's search for a venture capital firm to raise equity for a new project, should be that firm's familiarity with the parking industry and the probability of their entree to investors who may have a similar parking familiarity. It is not sufficient for a firm to have expertise in real estate financing or to have a coterie of real estate investors on hand; the educational process by a parking client would be time consuming and expensive.

FEES

The fees paid by a client owner for the services of a financial professional vary widely and are based on the level of service provided, the size of the financing and the type of relationship with the client. Such fees range from a low of $10,000 for a financial advisor to a municipality selling a small issue of general obligation bonds at public bidding, to as much as 8% of the total equity package arranged by a venture capital firm, possibly plus a portion of the equity for the firm's account.

There are no established standards for financial services fees. The fee structure is highly contingent upon the amount of work and analysis to be done by the financial professional. It is also dependent on the size of the total financing, as a unit percentage of the financing. As an example, more commissions are paid to salesmen in a publicly offered financing, debt or equity, than for a private sale to one or two institutional investors. Similarly, the level of sales commissions are higher for unrated securities, which are harder to sell, than for AAA rated securities, which are easier to sell.

While some financial advisory fees are paid by the client, whether or not the developed financial package is ever consummated, virtually all other fees for financial professional services are paid by the client only upon the successful completion of the transaction. Financial advisors, brokers and venture capital firms usually take their fee separately, upon a billing to the client. However, underwriters, merchant banks or commercial banks will take their fee as a discount off the price paid to the client for the purchase of the securities or mortgage.

The table in Appendix 14 shows a typical range of fees charged by financial professionals for different types of services and for various sizes of financing.

BIBLIOGRAPHY OF SOURCE MATERIAL

Guidelines for Insurance; published by Capital Guaranty Insurance Company, San Francisco, CA.

The Bond Buyer, October 2, 1989; published by American Banker Bond Buyer, New York, NY.

Memorandum of Mudge Rose Guthrie Alexander & Ferdon; July 28, 1989, Rule 15c2-12 under the Securities Exchange Act of 1934, New York, NY.

Tax Notes, October 31, 1988: Tax Exempt Financing in the Aftermath of the Technical and Miscellaneous Revenue Act of 1988; by Elena Marty-Nelson and Andrew Zack Blatter.

What Should the Consultant Do for His Client—the Consultant's Viewpoint; by J.L. Donoghue, Ralph H. Burke Associates, Park Ridge, Il; Reprinted from *Proceedings 1976 Workshop of the Institutional and Municipal Parking Congress,* Fredericksburg, VA.

The Use of Parking Consultants by Parking Organizations; published by the Institutional and Municipal Parking Congress, 1975, Fredericksburg, VA.

Credit Market Supply of Funds; published by the Federal Reserve Bank, June 2, 1989.

Investor Activity Reports, May 1989; published by the Security Industry Association, New York, NY.

Moody's Industrial Manual, 1989; published by Moody's Investor Service, New York, NY.

International Capital Adequacy Standards; Memorandum of June 22, 1988, by Paul Zoidis and Robin Maury, Shearson Lehman Hutton, New York, NY.

Memorandum of Mudge Rose Guthrie Alexander & Ferdon; June 1, 1989, Subject: Arbitrage Rebate Regulations, New York, NY.

Choosing Among Alternative Financing Structures: The Developer's Dilemma; by David C. Ling and Richard Peiser, published by Real Estate Review, Volume 17, No. 2, Summer 1987, Warren Gorham & Lamont Publications, Boston, MA.

Advantages of Participating Mortgages; by Peter Kelley, published by *Real Estate Review,* Volume 17, No. 2, Summer 1987, Warren Gorham & Lamont Publications, Boston, MA.

Financing Options; by E. Carlton Heeseler, prepared for the National Parking Association, Washington, D.C., Career Development Seminar Series, Section II, July 29, 1985.

The Impact of Ratings on Parking Bonds; by E. Carlton Heeseler, prepared for the Connecticut Association of Parking Agencies, Annual Conference, Darien, CT, October 27, 1976.

Security in a Bond Resolution, the Purpose of Various Provisions in a Bond Resolution; by E. Carlton Heeseler, prepared for the Connecticut Association of Parking Agencies, Annual Conference, Darien, CT, October 25, 1978.

Financing a Parking Garage; by E. Carlton Heeseler, prepared for the Northwest

Center for Professional Education, High Cash Flow Real Estate Niches Conference, Reston, VA, June 8, 1988.

Financing Public/Private Parking Facilities: A New Era Under Tax Reform for Non-Public Parking; by E. Carlton Heeseler, published by Parking Magazine, June 1986, National Parking Association, Washington, DC.

How to Finance a Parking Facility as a Public/Private Partnership; by E. Carlton Heeseler, prepared for the National Parking Association, Annual Convention, New York, NY, May 23, 1988.

Glossary of Municipal Securities Terms, First Edition, 1985; Published by the Municipal Securities Rulemaking Board, Washington, DC.

Distribution of Buyers, 30 Year Lease Revenue Financing; Targeting Market Segments; Certificates of Participation Transaction Structure; Lease Revenue Bond Transaction Structure; Comparative Interest Rates; prepared by Shearson Lehman Hutton Inc., July 1989.

Kenny Index, Historical Rates; published by Kenny Information Systems, Inc., August 1989, New York, NY.

Corporate Financial Policy, Third Edition, 1955; by Harry G. Guthmann and Herbert E. Dougall, published by Prentice-Hall, Inc., New York, NY.

FGIC Lease Criteria; prepared by Financial Guaranty Insurance Company, 1989, New York, NY.

Governmental Leasing: a Fifty State Survey of Legislation and Case Law; by George M. Mardikes, Pamela Cone, and Julie Van Horn, published in *The Urban Lawyer,* Volume 18, No. 1, Winter 1986, American Bar Association.

The Enterprise Capital Fund, an ADF National Pool; published by American Development Finance, Inc., Washington, DC, 1987.

Lease Purchase Financing Techniques; published by Shearson Lehman Hutton, Inc., 1989, New York, NY.

Guidelines for Using Parking Studies and Consultants; manuscript prepared by the Parking Consultants Council of the National Parking Association, April 24, 1989 by Robert A. Weant and Joseph McGee, Washington, DC.

Parking Facility Construction Payment Schedule Study; published by the National Parking Associates, 1984, Washington, DC, by E. Carlton Heeseler and William C. Arons.

APPENDIX 1

GLOSSARY

Some of the terms used in connection with tax exempt financing or real estate financing are unique to persons not familiar with the investment banking industry. The following glossary includes the definitions of selected terms as they are used in this study.

Ad Valorem Tax—A direct tax calculated "according to value" of property. Such tax is based on an assigned valuation (market or assessed) of real property and, in certain cases, on a valuation of tangible or intangible personal property. In virtually all jurisdictions the tax is a lien on the property enforceable by seizure and sale of property. An ad valorem tax is normally the one substantial tax which may be raised or lowered by a local governing body without the sanction of superior levels of government (although general restrictions, e.g., rate limitations, may exist on the exercise of this right for other than the payment of debt service); hence, ad valorem taxes often function as the balancing element in local budgets.

Amortization of Debt—The process of paying the principal amount of an issue of securities by periodic payments either directly to security holders or to a sinking fund for the benefit of securityholders.

Arbitrage—(1) Generally, transactions by which securities are bought and sold in different markets at the same time for the sake of the profit arising from a difference in prices in the two markets. (2) With respect to the issuance of municipal securities, arbitrage usually refers to the difference between the interest paid to the holders of the tax-exempt securities and the interest earned by investing the security proceeds in higher-yielding taxable securities. Internal Revenue Service regulations govern arbitrage on the proceeds from issuance of municipal securities. Investment of bond funds in securities yielding less than the interest rate on the bonds is called negative arbitrage.

Arbitrage Bonds—Bonds which are deemed by the Internal Revenue Service to violate Federal arbitrage regulations. If the Internal Revenue Service finds that bonds are "arbitrage bonds," the interest becomes taxable and therefore must be included in each bondholder's gross income for Federal income tax purposes.

Assessed Valuation—The appraised worth of property as set by a taxing authority for purposed of ad valorem taxation. It is important to note that the method of establishing assessed valuation varies from state to state, with the method generally specified by state law. For example, in certain jurisdictions the assessed valuation is equal to the full or market value of the property; in other jurisdictions the assessed valuation is equal to a set percentage of full or market value.

Authorizing Resolution or Ordinance—With respect to an issue of municipal securities the document adopted by the issuer which implements its power to issue the securities. The legal grant of such authority may be found in the enabling provisions of the constitution, statutes, charters and ordinances applicable to the issuer. Adoption of an authorizing resolution or ordinance by the issuer's governing body is a condition precedent to the issuance of the proposed securities.

Basis Point—1/100 of 1 percent of yield (0.01%). If a yield increases from 8.25% to 8.50%, the difference is referred to as a 25 basis point increase.

Blue-Sky Laws—A colloquial term for state securities laws, derived from a statement that such laws were directed at unethical promoters who "would sell building lots in the blue sky." Although these laws vary from state to state, most contain provisions concerning (a) prohibitions against fraud, (b) regulation of brokers and dealers doing business in the state, and (c) registration of securities. In many states certain of these requirements have been integrated with the federal securities laws, so that separate actions need not be taken to comply with them. Issues of municipal securities are generally exempt from state securities registration requirements, although brokers and dealers selling them are subject to many states' registration and regulatory requirements.

Bond—Evidence of the issuer's obligation to repay a specified principal amount on a date certain (maturity date), together with interest at a stated rate, or according to a formula for determining that rate. Bonds are distinguishable from notes, which usually mature in a much shorter period of time. Bonds may be classified according to maturity structure (serial vs. term), source of payment (general obligation vs. revenue), method of transfer (bearer vs. registered), issuer (state vs. municipality vs. special district) or price (discount vs. premium).

Bond Counsel—An attorney (or firm of attorneys) retained by the issuer to give a legal opinion that the issuer is authorized to issue the proposed securities, the issuer has met all legal requirements necessary for issuance, the bonds are valid and legally binding obligations of the issuer and interest on the proposed securities will be exempt from federal income taxation and, where applicable, from state and local taxation. Typically, bond counsel may prepare, or review and advise the issuer regarding authorizing resolutions or ordinances, trust indentures, official statements, validation proceedings and litigation. The bond counsel may also be referred to as the "bond attorney," the "bond approving attorney" or the "bond approving counsel."

Bond Election or Bond Referendum—A process whereby the voters of a governmental unit are given the opportunity to approve or disapprove a proposed issue of municipal securities. An election is most commonly required in connection with general obligation bonds. Requirements for voter approval may be imposed by constitution, statute or local ordinance.

Bond Fund—A colloquial term for a municipal securities investment company. The investment company holds a diversified portfolio of municipal securities, and units or shares in the investment company are sold to investors. There are two basic types of bond funds:

Unit Investment Trust or U.I.T.—A fixed portfolio of municipal securities sold to investors in trust "units," which represent fractional undivided ownership interests in the portfolio. The same securities are held in the portfolio until maturity or redemption. A U.I.T. is a "closed end" fund, i.e., with a fixed number of units available to investors.

Managed Fund—A fund, similar to a corporate securities mutual fund, in which investors own "shares" of a portfolio of municipal securities. An investment advisor actively manages the portfolio in line with the set investment policy; managed funds are also "open end" funds (i.e., they offer new shares to the public continuously). Managed funds may maintain portfolios of specialized types of municipal securities (e.g., short-term securities, or higher yielding long-term securities).

Bullet Loan—A debt obligation scheduled to be amortized over a long period (up to 30 years), but having a call feature at the option of the *lender.* Upon the exercise of such a call, the remaining principal balance outstanding becomes due and payable by the borrower.

Capitalized Interest or Funded Interest—A portion of the proceeds of an issue which is set aside to pay interest on the securities for a specified period of time. Interest is commonly capitalized for the construction period of a revenue-producing project. In an accounting sense capitalized interest is the interest expense paid during the construction period (net of income earned on construction funds) which is added to the book value of the asset being built.

Coverage or Debt Service Coverage—The ratio of pledged revenues available annually to pay the annual debt service requirement. Pledged revenues are usually calculated as net income before the deduction of interest, depreciation and amortization expenses. This ratio is one indication of the margin of safety for payment of debt service. The formula for determining coverage is as follows:

$$\text{Coverage} = \frac{\text{Pledge Revenues}}{\text{Debt Service Requirement}}$$

$$\text{Example: Coverage} = \frac{\$2{,}000{,}000}{\$1{,}200{,}000} = 1.66 \text{ times}$$

Debt Service Reserve Fund—The fund in which moneys are placed which may be used to pay debt service if pledged revenues are insufficient to satisfy the debt service requirements. The debt service reserve fund may be entirely funded with bond proceeds, or it may only be partly funded at the time of issuance and allowed to reach its full funding requirement over time, due to the accumulation of pledged revenues. If the debt service reserve fund is used in whole or part to pay debt service, the issuer usually is required to replenish the funds from the first available funds or revenues. A typical reserve requirement might be the maximum aggregate annual debt service requirement for any year remaining until the bonds reach maturity. The size and investment of the reserve may be subject to arbitrage regulations. Under a typical revenue pledge this fund is the third to be funded out of the revenue fund.

Enterprise Activity—A revenue-generating project or business which provides net income to pay debt service on securities issued to finance the facility. The debts of such projects are self-liquidating when the projects earn sufficient moneys to cover all debt service and other requirements imposed under the bond contract. Common examples include water and sewer plants, electric supply facilities, parking systems and private business projects financed with industrial development bonds.

Financial Advisor—With respect to a new issue of municipal securities, a consultant who advises the issuer on matters pertinent to the issue, such as structure, timing, marketing, fairness of pricing, terms and bond ratings. A financial advisor may also be employed to provide advice on subjects unrelated to a new issue of municipal securities, such as advising on cash flow and investment matters. The financial advisor is sometimes referred to as a "fiscal consultant" or "fiscal agent." MSRB Rule G-23 provides that a firm or bank which has acted in a financial advisory capacity with respect to a new issue of municipal securities (pursuant to a written contract) may underwrite the new issue (a) on a negotiated basis after making certain disclosures, obtaining the consent of the issuer and terminating the financial advisory relationship; or (b) on a competitive basis if the issuer gives written consent before the financial advisor's bid is submitted.

Full Disclosure—The principle that accurate and complete information material to a securities transaction which a potential investor would be likely to consider important in making investment decisions must be made available to purchasers or prospective purchasers. Material facts may include descriptions of the issuer and the true obligor in a conduit financing, as well as the structure of a bond issue and the security therefor. Full disclosure enables the investors to evaluate the credit quality of an issue. The material facts pertinent to a new issue of municipal securities are generally disclosed in the official statement, which also states that there have been no material misstatements, or omissions by the issuer with respect to the issue, and that no facts have become known which would render false or misleading any statement which was made. Unlike full disclosure requirements in connection with corporate debt and equity securities, filing, registration and reporting to the Securities and Exchange Commission are not required with respect to municipal securities.

General Obligation Bond or G.O. Bond—A bond which is secured by the full faith and credit of an issuer with taxing power. General obligation bonds issued by local units of government are typically secured by a pledge of the issuer's ad valorem taxing power; general obligation bonds issued by states are generally based upon appropriations made by the state legislature for the purposes specified. Ad valorem taxes necessary to pay debt service on general obligation bonds are often not subject to any constitutional property tax millage limits. Such bonds constitute debts of the issuer and normally require approval by election prior to issuance. In the event of default, the holders of general obligation bonds have the right to compel a tax levy or legislative appropriation, by mandamus or injunction, in order to satisfy the issuer's obligation on the defaulted bonds.

Industrial Development Bond or IDB—In general, securities issued by a state, a local government or development agency to finance the construction or purchase of industrial, commercial or manufacturing facilities to be purchased by or leased to a private user. IDBs are backed by the credit of the private user and generally are not considered liabilities of the governmental issuer (although in some jurisdictions they may also be backed by an issuer with taxing power). While the authorization to issue IDBs is provided by a state statute, the tax-exempt status of these bonds is derived from federal law (I.R.C. §103(b)(2)). IDBs are also referred to as industrial revenue bonds (IRBs).

Investment Grade—The broad credit designation given bonds which have a high probability or being paid and minor, if any, speculative features. Bonds rated BBB or higher by Standard and Poor's Corporation or Baa or higher by Moody's Investors Service, Inc. are deemed by those agencies to be "investment grade." Banking law requires that certain types of securities acquired by bank portfolios must be both marketable and "investment grade"; it should be noted that the presence (or absence) of a rating deemed "investment grade" by the rating agencies is not necessarily indicative of a security's status for purposes of banking law.

Legal Opinion or Approving Opinion—The written conclusions of law, as stated by bond counsel to the effect that the issuance of municipal securities and the proceedings taken in connection therewith comply with applicable laws and that such securities are valid and binding obligations of the issuer and that interest on the securities will be exempt from federal income taxation and, where applicable, from state and local taxation. The legal opinion is generally printed on the reverse side of the securities.

Letter of Credit or LC or L.O.C.—A commitment, usually made by a commercial bank, to honor demands for payment of a debt upon compliance with conditions and/or the occurrence of certain events specified under the terms of the commitment. In municipal financings, bank letters of credit are sometimes used as sources of security enhancement for issues of municipal notes, commercial paper or bonds, with the bank issuing the letter of credit committing to pay the principal of and interest on the securities in the event that the issuer is unable to do so. A letter of credit may also be used to provide liquidity for the retirement of commercial paper or put bonds.

Level Debt Service—A maturity schedule in which the combined annual amount of principal and interest payments remains relatively constant over the life of the issue; tax exempt bonds predominantly are issued in minimum $5,000 denominations. (applicable provisions of law of the individual states may impact upon the implementation of level debt service).

EXAMPLE:

Level Debt Service Assumptions:
Size of issue: $10,000,000
Interest rate: 7%
Maturity of issue: 5 years

Debt Schedule

Years	*Principal*	*Interest*	*Total**
1	$ 1,740,000	$ 700,000	$ 2,440,000
2	1,860,000	578,200	2,438,200
3	1,990,000	448,000	2,438,000
4	2,130,000	308,700	2,438,700
5	2,280,000	159,600	2,439,600
TOTAL	$10,000,000	$2,194,500	$12,194,500

*Total of principal and interest remains substantially level throughout life of issue.

Leverage—The ratio of debt to the total amount of financing of a project, i.e., debt plus equity. This ratio is usually 100% for municipality owned financing. Private taxable financing can vary widely from as low as 0% (all equity) to as high as 100% for options such as participating mortgages. Commercial bank conventional financing normally requires a leverage of 70-80%.

Loan to Value Ratio—The ratio of the amount of debt borrowed to finance a project to the appraised value of the land and improvements. The lower this ratio the more advantageous to a developer, although with some highly leveraged financings, this loan to value ratio can be as high as 90%.

Marketability—The ease or difficulty with which securities can be sold in the market. An issue's marketability depends upon many factors, including its interest rate, security provisions, maturity and credit quality, plus (in the case of the sale of a new issue) the size of the issue, the timing of its issuance, and the volume of comparable issues being sold.

Municipal Securities—A general term referring to securities issued by local governmental subdivisions such as cities, towns, villages, counties, authorities, or special districts, as well as securities issued by states and political subdivisions or agencies of states. A prime feature of these securities is that interest on them is generally exempt from federal income taxation and income taxation of the issuing state and its local taxing jurisdictions.

Municipal Securities Rulemaking Board or MSRB—an independent self-regulatory organization established by the Securities Acts Amendments of 1975, which is charged with primary rulemaking authority over dealers, dealer banks and brokers in municipal securities. Its 15 members represent three categories—securities firms, bank dealers and the public—each category having equal representation on the Board.

Negotiated Sale—The sale of a new issue of securities by an issuer through an exclusive agreement with an underwriter or underwriting syndicate selected by the issuer. A negotiated sale should be distinguished from a competitive sale, which requires public bidding by the underwriters. The primary points of negotiation for an issuer are the interest rate on debt and purchase price of the debt or equity. The sale of a new issue of securities in this manner is also known as a negotiated underwriting.

Net Interest Cost or NIC—A common method of computing the interest expense to the issuer of debt, which usually serves as the basis of award in a competitive sale. NIC takes into account any premium and discount paid on the issue. NIC represents the dollar amount of coupon interest payable over the life of the debt, without taking into account the time value of money (as would be done in other calculation methods, such as the "true interest cost" method). While the term "net interest cost" actually refers to the dollar amount of the issuer's interest cost, it is also used to refer to the overall rate of interest to be paid by the issuer over the life of the bonds. The formula for calculating the NIC rate is:

$$\frac{\text{Total Coupon Interest Payments} + \text{Discount (or} - \text{Premium)}}{\text{Bond Years}} = \text{NIC}$$

Note—A written, short-term promise of an issuer to repay a specified principal amount on a date certain, together with interest at a stated rate, payable from a defined source of anticipated revenue. Notes usually mature in one year or less, although notes of longer maturities are also issued, subject to applicable provisions of law. The following types of notes are common in the municipal market:

- Bond Anticipation Notes (BANs)—Notes issued by a governmental unit, usually for capital projects, which are paid from the proceeds of the issuance of long term bonds.
- Grant Anticipation Notes (GANs)—Notes issued on the expectation of receiving grant moneys, usually from the federal government. The notes are payable from the grant funds, when received.
- Revenue Anticipation Notes (RANs)—Notes issued in anticipation of receiving revenues at a future date.
- Tax Anticipation Notes (TANs)—Notes issued in anticipation of future tax receipts, such as receipts of ad valorem taxes which are due and payable at a set time of the year.

GANs, RANs, and TANs may be issued to finance capital projects or to alleviate cash flow problems of the issuer.

Over-the-Counter Market—A market for securities which are traded other than on a national securities exchange. The over-the-counter (or OTC) market is characterized most particularly by a system of dealer market-making rather than the auction market system common on the securities exchanges. Almost all municipal securities are traded exclusively in the OTC market.

Par Value—The amount of principal which must be paid at maturity. The par value is also referred to as the face amount of a security.

Paying Agent—The entity responsible for transmitting payments of interest and principal from an issuer of municipal securities to the security holders. The paying agent is usually a bank or trust company, but may be the treasurer or some other officer of a municipal issuer. The paying agent may also provide other services for the issuer such as reconciliation of the securities and coupons paid, destruction of paid securities and coupons, and similar services.

Pledged Revenues—The moneys obligated for the payment of debt service and other deposits required by the bond contract.

- Gross Pledge or Gross Revenue Pledge—A pledge that all revenues received will be used for debt service prior to deductions for any costs or operating expenses.
- Net Pledge or Net Revenue Pledge—A pledge that all funds remaining after certain operational and maintenance costs and expenses are paid, will be used for payment of debt service.

Preliminary Official Statement or Red Herring or P.O.S.—A preliminary version of the official statement which is used by an issuer or underwriters to describe the proposed issue of municipal securities prior to the determination of the interest rate(s) and offering price(s). The preliminary official statement may be used by issuers to gauge underwriters' interest in an issue and is often relied upon by potential purchasers in making their investment decisions. Normally, offers for the sale of or acceptance of securities are not made on the basis of the preliminary official statement, and a statement to that effect appears on the face of the document generally in red print, which gives the document its nickname, *red herring*. The preliminary official statement is technically a draft.

Private Placement—With respect to municipal securities, a negotiated sale in which the new issue securities are sold directly to institutional or private investors rather than through a public offering. Issuers often require investors purchasing privately placed securities to agree to restrictions as to resale; the investor may provide a signed agreement to abide by those restrictions.

Public Authority—A unit or agency of government usually established by the state government to perform specialized functions, usually financed by service charges, fees or tolls, although it may also have taxing powers. In many cases authorities have the power to issue debt which is secured by the lease rental payments made to the Authority by a governmental unit using the facilities constructed with bond proceeds. An authority may function independently of other governmental units, or it may depend upon other units for its creation, funding or administrative oversight. Examples of authorities include health facilities authorities, industrial development authorities, parking authorities and housing authorities.

Ratings—Evaluations of the credit quality of notes and bonds made by independent rating services. Ratings are intended to measure the probability of the timely repayment of principal of and interest on municipal securities. Ratings are initially made before issuance and are periodically reviewed and may be amended to reflect changes in the issuer's credit position. The information required by the rating agencies varies with each issue, but generally includes information regarding the issuer's demographics, debt burden, economic base, finances and management structure. Many financial institutions also assign their own individual ratings to securities.

Redemption Provisions or Call Features—The terms of the bond contract giving the *issuer* the right or requiring the issuer to redeem or "call" all or a portion of an outstanding issue of bonds prior to their stated dates of maturity at a specified price, usually at or above par. Common types of redemption provisions include:

- Optional Redemption—The issuer has the right to redeem bonds, usually after a stated date and at a premium, but is not required to do so.
- Mandatory Redemption—The issuer is required to call outstanding bonds based on a predetermined schedule or as otherwise provided in the bond contract. Mandatory redemption provisions are included with the creation of a sinking fund for term maturities. Frequently, the issuer is allowed to make open market purchases in lieu of calling the bonds.
- Extraordinary Optional Redemption—The issuer has the right to call or redeem an issue of bonds upon the occurrence of certain events. For example, the right to extraordinary optional redemption of an issue of bonds may be exercised when mortgages are prepaid in connection with a housing revenue bond issue.
- Extraordinary Mandatory Redemption—The issuer is required to call or redeem all or part of an issue of bonds upon the occurrence of certain events. For example, the issuer may be required to call or redeem bonds when proceeds of an issue are not expended for the purpose of the issue as of a given time; when excess bond proceeds exist after completion of a project; or from insurance proceeds when the facility has been substantially destroyed due to an accident. The latter situation is also known as a "calamity call" or "catastrophe call."

Refunding—A procedure whereby an issuer refinances an outstanding bond issue by issuing new bonds. There are generally two major reasons for refunding: to reduce the issuer's interest costs or to remove a burdensome or restrictive covenant imposed by the terms of the bonds being refinanced. The proceeds of the new bonds are either deposited in escrow to pay the debt service on the outstanding obligations, when due (in which case the financing is known as an "advance refunding"), or used to immediately retire the outstanding obligations. The new obligations are referred to as the "refunding bonds," and the outstanding obligations being refinanced are referred to as the "refunded bonds" or the "prior issue." For accounting purposes, refunded obligations are not considered a part of the issuer's debt because the lien of the holders of the refunded bonds, in the first instance, is on the escrowed funds, not on the originally pledged source of revenues. The refunded bonds, however, will continue to hold a lien on the originally pledged source of revenues unless provisions have been made in the bond contract on the refunded bonds for defeasance of the bonds prior to redemption.

Registered Bond—A bond whose owner is designated on records maintained for this purpose by a registrar, the ownership of which cannot be transferred without the registrar recording the transfer on these records. The principal and interest on "fully registered" bonds is paid directly by check (or other funds transfer) to the registered owner. Bonds issued prior to July 1983 may sometimes be "registered as to principal only"; in this form coupons reflecting the interest payments due remain attached to the bonds and must be detached and redeemed in order to receive the interest due. Most municipal securities issued after June 1983 are in "fully registered" form due to provisions of the Internal Revenue Code which deny tax-exemption to interest paid on issues (other than those exempt from this provision) which are not in registered form.

Revenue Bond—A bond which is payable from specific source of revenue and to which the full faith and credit of an issuer with taxing power is not pledged. Revenue bonds are payable from identified sources of revenue, and do not permit the bondholders to compel taxation or legislative appropriation of funds not pledged for payment of debt service. Pledged revenues may be derived from operation of the financed project or other specified non-ad valorem taxes. Generally, no voter approval is required prior to issuance of such obligations.

Special Assessment—A charge imposed against property in a particular locality because that property receives a special benefit, by virtue of some public improvement, separate and apart from the general benefit accruing to the public at large. Special assessments must be apportioned according to the value of the benefit received, rather than the cost of the improvement, and may not exceed the value of such benefit or the cost of the improvement, whichever is less.

Special Districts—Single-purpose or limited-purpose units of government formed under state enabling legislation to meet certain local needs not satisfied by existing general purpose governments in a given geographic area. In some states, special districts (such as school districts) may be granted taxing powers.

Special Tax Bond—A bond secured by one or more designated taxes other than ad valorem taxes. For example, bonds for a particular purpose might be supported by sales, cigarette, fuel or business license taxes; however, the designated tax does not have to be directly related to the project purpose. Such bonds are not considered "self-supporting" debt.

Spread—With respect to new issue municipal securities, the differential between the price paid to the issuer for the new issue and the prices at which the securities are initially offered to the investing public; this is also termed the "gross spread" or "gross underwriting spread." To the extent that the initial offering prices are subsequently lowered by the syndicate, the full amount of the spread may not be realized by the syndicate. The spread is usually expressed in points or fractions thereof. The spread generally has four components:

- Expenses—The costs of operating the syndicate for which the lead manager may be reimbursed.
- Management Fee—The amount paid to the lead manager for handling the affairs of the syndicate and initial development of the financing.
- Takedown—Normally the largest component of the spread, similar to a commission, which represents the income derived from the sale of the securities. If bonds are sold by a member of the syndicate, the seller is entitled to the full takedown (also called the "total takedown"); if bonds are sold by a dealer which

is not a member of the syndicate, such seller receives only that portion of the takedown known as the concession or dealer's allowance, with the balance (often termed the "additional takedown") retained by the syndicate. Each seller of bonds pays a portion of the take down received to the salesman consummating the sale to an investor.

- Underwriting Risk—The amount of profit or spread left in a syndicate account after meeting all other expenses or deductions. A portion of the residual is paid to each underwriter within a syndicate on a pro rata basis according to the number of bonds each dealer has committed to sell without regard to the actual sales by each member.

Syndicate—A group of underwriters formed to purchase (underwrite) a new issue of municipal securities from the issuer and offer it for resale to the general public. The syndicate is organized for the purposes of sharing the risks of underwriting the issue, obtaining sufficient capital to purchase an issue and for broader distribution of the issue to the investing public. One of the underwriting firms will be designated as the syndicate manager or lead manager to administer the operations of the syndicate. There are two major types of syndicate agreements:

- Divided or Western Account—A method for determining liability stated in the agreement among underwriters in which each member of an underwriting syndicate is liable only for the amount of its participation in the issue, and not for any unsold portion of the participation amounts allocated to the other underwriters.
- Undivided or Eastern Account—A method for determining liability stated in the underwriting agreement in which each member of the underwriting syndicate is liable for any unsold portion of the issue according to each member's percentage participation in the syndicate.

Tax-Exempt Bond—Another term for a municipal bond. Interest on municipal securities is exempt from federal income taxation pursuant to Section 103 of the Internal Revenue Code, and may or may not be exempt from state income or personal property taxation in the jurisdiction where issued. If the bond is exempt from state income tax, it possesses "double exemption" status. "Triple exemption" bonds are exempt from municipal or local income taxes, as well as from federal and state income tax.

True Interest Cost or TIC—Also known as Canadian Interest Cost. Under this method of computing, the borrower's cost is defined as the rate, compounded semi-annually, necessary to discount the amounts payable on the respective principal and interest payment dates to the purchase price received for the new issue securities. TIC computations produce a figure slightly different from the net interest cost (NIC) method since TIC considers the time value of money, while NIC does not.

Trust Indenture—A contract between the issuer of municipal securities and a trustee, for the benefit of the bondholders. The trustee administers the funds or property specified in the indenture in a fiduciary capacity on behalf of the bondholders. The trust indenture, which is generally a part of the bond contract, establishes the rights, duties, responsibilities and remedies of the issuer and trustee and determines the exact nature of the security for the bonds. The trustee is generally empowered to enforce the bond contract on behalf of the bondholders.

Zero Coupon Bond—An original issue discount bond on which no periodic interest payments are made, but which is issued at a deep discount from par, accreting (at the rate represented by the offering yield at issuance) to its full value at maturity.

Appendix 2. Summary of Private Domestic Holdings[1] (Billions of Dollars)

FIRST QUARTER YEAR	U.S. GOVERNMENT SECURITIES		TAX-EXEMPT OBLIGATIONS		CORPORATE AND FOREIGN BONDS		RESIDENTIAL MORTGAGES		OTHER MORTGAGES AND LOANS[2]		OTHER EQUITIES		TOTAL PRIVATE HOLDINGS
	AMOUNT	%	AMOUNT	%	AMOUNT	%	AMOUNT	%	AMOUNT	%	AMOUNT	%	
1982	1,064.9	17.6	417.9	6.9	427.1	7.1	934.1	15.4	1,491.5	24.6	1,720.9	28.4	6,056.4
1983	1,290.4	18.8	471.7	6.9	441.7	6.4	992.2	14.4	1,649.6	24.1	2,021.9	29.4	6,867.5
1984	1,524.9	20.2	522.1	6.9	476.8	6.3	1,096.5	14.5	1,915.2	25.3	2,021.5	26.8	7,557.0
1985	1,805.2	20.3	658.4	7.4	517.6	5.8	1,185.1	13.3	2,129.2	24.0	2,584.3	29.2	8,879.8
1986	2,162.6	21.7	689.2	6.9	601.7	6.0	1,254.7	12.6	2,328.1	23.3	2,948.5	29.5	9,984.8
1987	2,422.4	23.0	723.7	6.9	688.1	6.5	1,351.1	12.8	2,510.2	23.8	2,858.3	27.0	10,553.8
1988	2,632.6	23.0	756.8	6.6	769.1	6.7	1,480.8	12.9	2,705.4	23.6	3,122.6	27.2	11,467.3
1989	2,697.6	23.0	764.9	6.5	782.1	6.7	1,491.9	12.7	2,750.2	23.4	3,246.0	27.7	11,732.7
1990	2,962.8	23.1	822.4	6.4	847.5	6.6	1,680.1	13.1	2,922.4	22.7	3,620.0	28.1	12,855.2

[1]Excludes holdings of Federal agencies, foregin holders, financial intermediation, and deposits and currency.
[2]Includes farm and commercial mortgages.
Excludes mutual fund shares.

Source: Federal Reserve Bank, Credit Market Supply of Funds, December 1990.

Appendix 3. Asset Holdings of Selected Securities 1989, First Quarter (Billions of Dollars)

DESCRIPTION OF HOLDERS	CORPORATE AND FOREIGN BONDS		CORPORATE EQUITY AND MUTUAL FUNDS		COMMERCIAL MORTGAGES		TAX EXEMPT SECURITIES		TOTAL CORPORATE AND COMMERCIAL		TOTAL SELECTED SECURITIES	
	AMOUNT	%	AMOUNT	%	AMOUNT	%	AMOUNT	%	AMOUNT	%	AMOUNT	%
Households	80.5	5.9	2,282.3	61.2	12.4	1.7	272.4	35.6	2,375.2	40.9	2,647.6	40.3
Nonfinancial Corporate Business	—	—	18.1	0.5	25.9	3.6	9.6	1.3	44.0	0.8	53.6	0.8
State and Local Governments	158.3	11.6	238.2	6.4	12.9	1.8	10.6	1.4	409.4	7.0	420.0	6.4
Commercial Banks	78.6	5.8	—	—	317.3	44.4	149.9	19.6	395.9	6.8	545.8	8.3
Savings & Loan Associations	54.0	4.0	—	—	108.3	15.2	1.3	0.2	162.3	2.8	163.6	2.5
Mutual Savings Banks	13.3	1.0	8.5	0.2	29.3	4.2	1.9	0.2	51.1	0.9	53.0	0.8
Money Market Funds	—	—	—	—	—	—	67.6	8.8	—	—	67.6	1.0
Mutual Funds	67.4	4.9	194.3	5.2	—	—	82.4	10.8	261.7	4.5	344.1	5.2
Life Insurance Companies	445.3	32.6	116.8	3.2	185.8	26.0	11.0	1.4	747.9	12.8	758.9	11.5
Private Pension Funds	185.9	13.6	577.2	15.3	3.9	0.5	—	—	767.0	13.2	767.0	11.7
Other Insurance Companies	64.3	4.7	79.0	2.2	4.6	0.6	156.7	20.5	147.9	2.5	304.6	4.6
Brokers and Dealers	27.5	2.0	12.2	0.3	—	—	1.5	0.2	39.7	0.7	41.2	0.6
Foreign Investors	190.1	13.9	205.8	5.5	—	—	—	—	395.9	6.8	395.9	6.0
Finance Companies	—	—	—	—	4.4	0.6	—	—	4.4	0.1	4.4	0.1
REITs	—	—	—	—	5.2	0.7	—	—	5.2	0.1	5.2	0.1
U.S. Government	—	—	—	—	5.1	0.7	—	—	5.1	0.1	5.1	0.1
Total Holdings	1,365.2	100.0	3,732.4	100.0	715.1	100.0	764.9	100.0	5,812.7	100.0	6,577.6	100.0

Source: Federal Reserve Bank, Private Securities, June 2, 1989.

Appendix 4. Holdings of Tax Free Investments

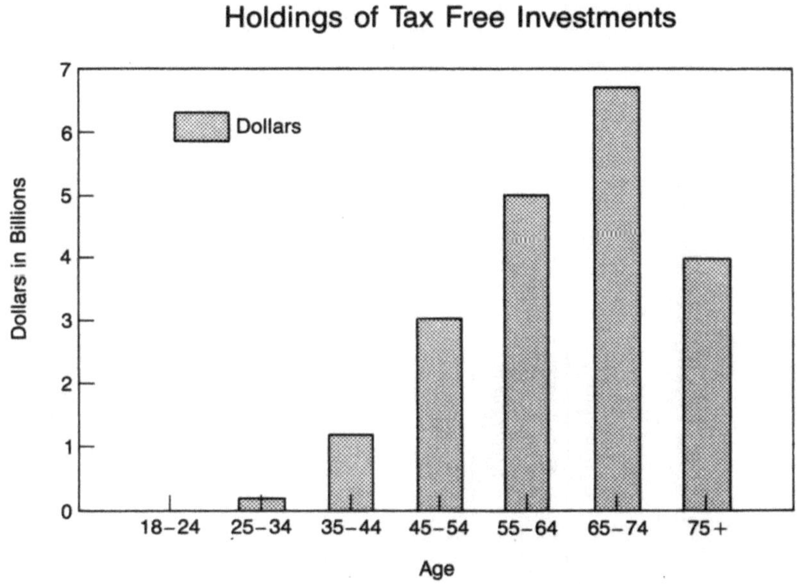

Appendix 5. Public Securities Association Trends in the Holdings of Municipal Securities (Billions of Dollars)

YEAR	TOTAL AMOUNT OUT-STANDING ($ BILLIONS)	HOUSEHOLDS $ AMOUNT OUTSTANDING*	HOUSEHOLDS % OF TOTAL OUTSTANDING*	MUTUAL FUNDS $ AMOUNT OUTSTANDING	MUTUAL FUNDS % OF TOTAL OUTSTANDING	MONEY MARKET FUNDS $ AMOUNT OUTSTANDING	MONEY MARKET FUNDS % OF TOTAL OUTSTANDING	COMMERCIAL BANKS $ AMOUNT OUTSTANDING	COMMERCIAL BANKS % OF TOTAL OUTSTANDING	PROPERTY AND CASUALTY INSURANCE COMPANIES $ AMOUNT OUTSTANDING	PROPERTY AND CASUALTY INSURANCE COMPANIES % OF TOTAL OUTSTANDING	OTHER $ AMOUNT OUTSTANDING	OTHER % OF TOTAL OUTSTANDING
1960	70.8	30.8	43.5	—	—	—	—	17.7	25.0	8.1	11.4	14.2	20.1
1961	75.9	32.0	42.2	—	—	—	—	20.5	27.0	9.1	11.9	14.3	18.9
1962	81.2	31.5	38.8	—	—	—	—	25.7	31.6	9.9	12.2	14.2	17.4
1963	86.9	32.0	36.8	—	—	—	—	30.1	34.6	10.6	12.2	14.2	16.4
1964	92.9	34.6	37.3	—	—	—	—	33.7	36.2	11.0	11.8	13.7	14.7
1965	100.3	36.4	36.3	—	—	—	—	38.8	38.7	11.3	11.3	13.7	13.7
1966	105.9	40.4	38.2	—	—	—	—	41.2	38.9	12.0	11.4	12.2	11.5
1967	113.7	38.1	33.5	—	—	—	—	50.3	44.3	13.5	11.9	11.7	10.3
1968	123.2	37.6	30.5	—	—	—	—	58.9	47.8	14.4	11.7	12.3	10.0
1969	133.1	46.9	35.2	—	—	—	—	59.5	44.7	15.5	11.6	11.2	8.4
1970	144.4	46.0	31.9	—	—	—	—	70.2	48.6	17.0	11.8	11.1	7.7
1971	161.8	46.1	28.5	—	—	—	—	82.8	51.2	20.5	12.7	12.4	7.7
1972	176.5	48.4	27.4	—	—	—	—	90.0	51.0	24.8	14.1	13.4	7.6

1973	191.2	53.7	28.1	—	—	—	—	96.7	50.5	28.5	14.9	12.4	6.5
1974	207.7	61.9	29.8	—	—	—	—	101.1	48.7	30.7	14.8	14.0	6.8
1975	223.8	68.1	30.4	—	—	—	—	102.9	46.0	33.3	14.9	19.6	8.7
1976	239.5	70.1	29.3	0.5	0.2	—	—	106.0	44.2	38.7	16.1	24.3	10.1
1977	261.4	68.6	26.2	2.2	0.8	—	—	115.2	44.1	49.4	18.9	26.1	10.0
1978	289.7	71.2	24.6	2.7	0.9	—	—	126.2	43.6	62.9	21.7	26.7	9.2
1979	320.1	81.7	25.5	4.0	1.3	—	—	135.6	42.4	72.8	22.7	25.9	8.1
1980	350.3	88.3	25.2	4.4	1.3	1.9	0.5	149.2	42.6	80.5	23.0	26.0	7.4
1981	373.7	99.8	26.7	5.1	1.4	4.2	1.1	154.2	41.3	83.9	22.5	26.5	7.1
1982	417.9	120.2	28.8	8.0	1.9	13.2	3.2	158.5	37.9	87.0	20.8	31.1	7.4
1983	469.0	155.1	33.1	13.4	2.9	16.8	3.6	163.2	34.8	86.7	18.5	33.8	7.2
1984	520.0	180.7	34.8	19.1	3.7	23.8	4.6	174.6	33.6	84.7	16.3	37.1	7.1
1985	655.5	218.8	33.4	33.5	5.1	36.3	5.5	231.7	35.3	88.2	13.5	47.0	7.2
1986	679.1	197.1	29.0	65.3	9.6	63.8	9.4	203.4	30.0	101.9	15.0	47.6	7.0
1987	728.4	255.9	35.1	70.7	9.7	61.4	8.4	174.3	23.9	124.8	17.1	41.3	5.7
1988	790.0	319.2	40.4	78.7	10.0	65.7	8.3	151.6	19.2	134.1	17.0	40.7	5.2
1989	821.2	348.5	42.4	93.6	11.4	69.4	8.5	133.8	16.3	135.9	16.5	40.0	4.9
1990 Q3	$836.3	$347.5	41.6	$104.2	12.5	$79.0	9.4	$122.7	14.7	$136.6	16.3	$46.3	5.5

*Includes Unit Trusts

Source: Flow of Funds Accounts, Third Quarter 1990.
Flow of Funds Accounts Financial Assets and Liabilities Year-end, 1961-1984.
Board of Governors of the Federal Reserve System, Washington, D.C.

Appendix 6. Holders of Long Term Municipal Debt: 1978-1990

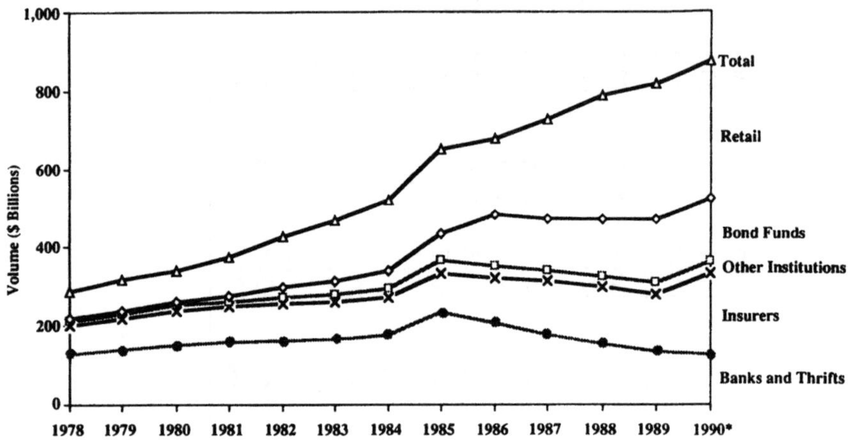

* Adjusted figures for first 6 months only.
Source. Federal Reserve Board, "Flow of Funds Accounts."

Appendix 7. Purposes of Municipal Financing

January-December					
	1990		1989		
	Volume	Mkt Share	Volume	Mkt Share	Pct. Chg
TOTAL	$126,373,500	100.0%	$125,035,800	100.0%	1.1%
First Quarter	25,981,600	20.6	25,485,000	20.4	+1.9
Second Quarter	35,384,100	28.0	31,292,800	25.0	+13.1
Third Quarter	33,398,800	26.4	32,002,700	25.6	+4.4
Fourth Quarter	31,608,900	25.0	36,255,200	29.0	-12.8
Tax-Exempt	105,143,000	83.2	104,760,300	83.8	+0.4
Minimum-Tax	18,381,500	14.5	16,898,700	13.5	+8.8
Taxable	2,849,000	2.3	3,376,800	2.7	-15.6
New-Money	105,006,100	83.1	89,940,400	71.9	+16.8
Refunding	21,367,400	16.9	35,095,400	28.1	-39.1
Negotiated	93,851,600	74.3	92,319,800	73.8	+1.7
Competitive	30,408,400	24.1	29,927,800	23.9	+1.6
Private Placement	2,113,500	1.7	2,788,200	2.2	-24.2
Revenue	86,444,300	68.4	86,818,900	69.4	-0.4
General Obligation	39,929,200	31.6	38,216,900	30.6	+4.5
Fixed-Rate	114,209,353	90.4	112,924,539	90.3	+1.1
Variable-Rate	12,164,147	9.6	12,111,261	9.7	+0.4
Bond Insurance	32,890,778	26.0	30,996,796	24.8	+6.1
Letters of Credit	9,941,600	7.9	10,095,400	8.1	-1.5
Insured Mortgage	4,263,600	3.4	4,435,300	3.5	-3.9
Surety Bond	149,900	0.1	61,100	0.0	+145.3
Education	21,705,600	17.2	19,055,400	15.2	+13.9
Electric Power	5,153,900	4.1	7,973,800	6.4	-35.4
Health Care	13,591,900	10.8	15,807,900	12.6	-14.0
Housing	15,398,500	12.2	11,875,300	9.5	+29.7
Industrial Development	4,310,000	3.4	5,648,400	4.5	-23.7
Public Facilities	5,685,500	4.5	4,933,900	3.9	+15.2
Solid Waste/Pollution	5,059,500	4.0	4,957,700	4.0	+2.1
Transportation	13,235,600	10.5	10,834,500	8.7	+22.2
Utilities	10,381,300	8.2	12,358,300	9.9	-16.0
Other	31,851,700	25.2	31,590,600	25.3	+0.8

Appendix 7. *(Continued)*

January-December

	1990		1989		
	Volume	Mkt Share	Volume	Mkt Share	Pct. Chg
State Governments	15,469,500	12.2	12,122,100	9.7	+27.6
State Agencies	41,128,000	32.5	39,175,400	31.3	+5.0
Municipalities........................	47,050,450	37.2	47,890,100	38.3	-1.8
Local Authorities..................	21,195,500	16.8	23,868,600	19.1	-11.2
Universities & Colleges	1,529,200	1.2	1,971,200	1.6	-22.4

NOTES: Dollar amounts are in thousands. Private placements are included, but short-term notes and remarketings are excluded.

Source: Securities Data Co./Bond Buyer (1/1/91)

Appendix 8. Certificates of Participation Transaction Structure

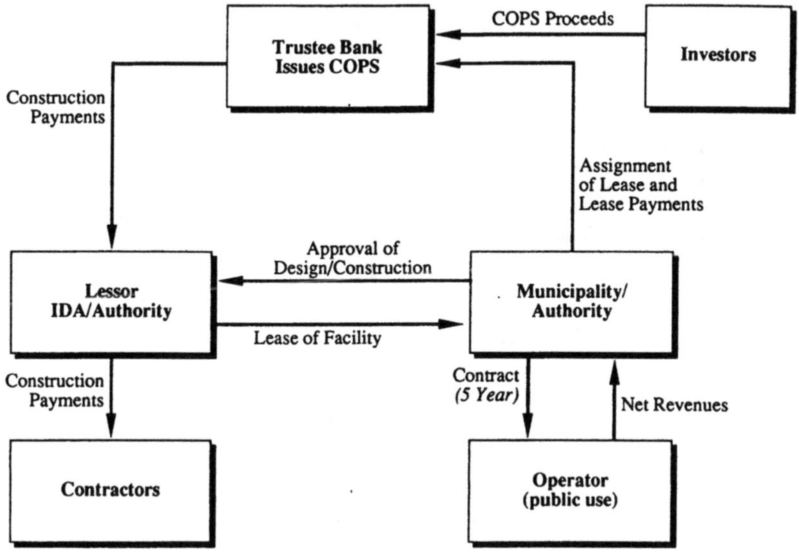

Appendix 9. Lease Revenue Bond Transaction Structure

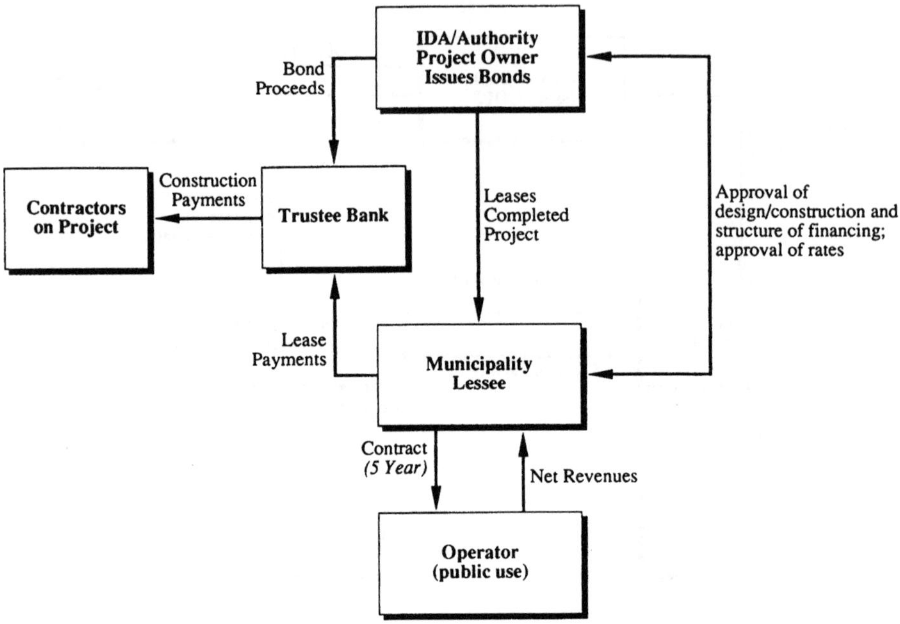

Appendix 10. Comparative Interest Rates. 20-Yr. Bond Buyer G.O. vs. Corporate "A" Rated Utility vs. 30-yr. Treas

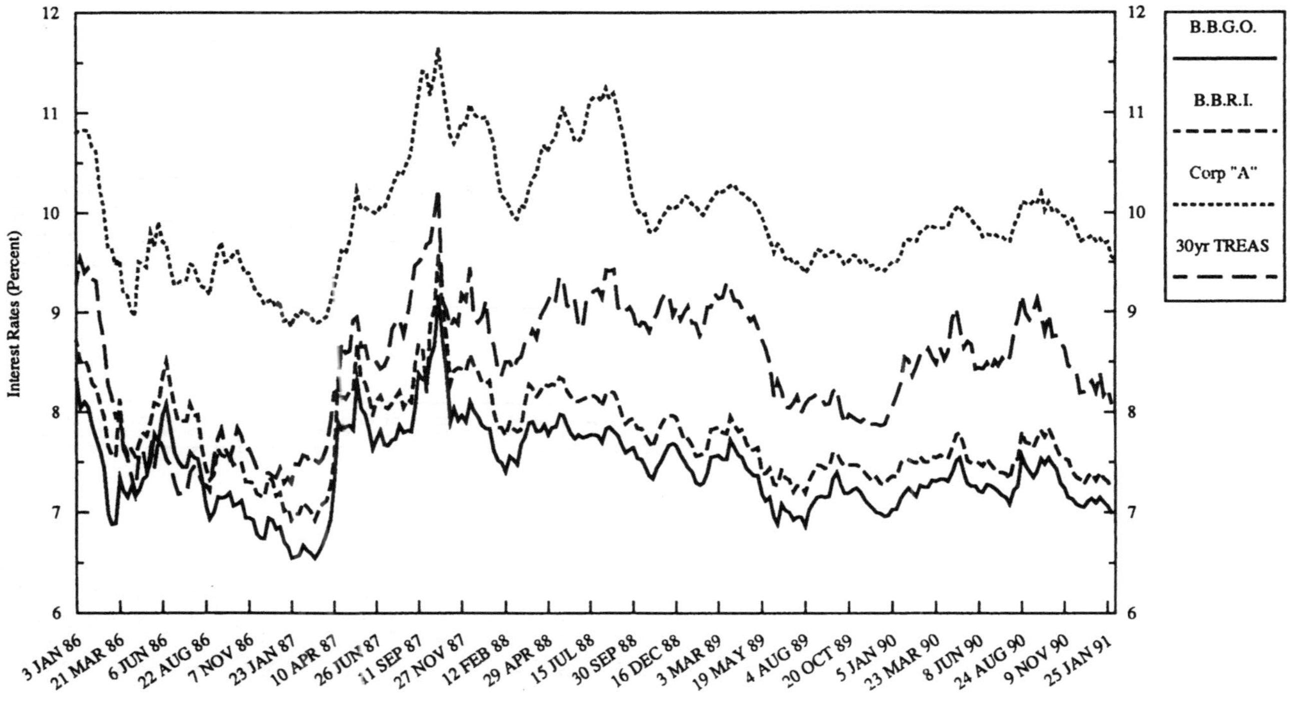

Weekly: 1/1/86 to 2/6/91

Appendix 11. Kenny Index—Historical Rates

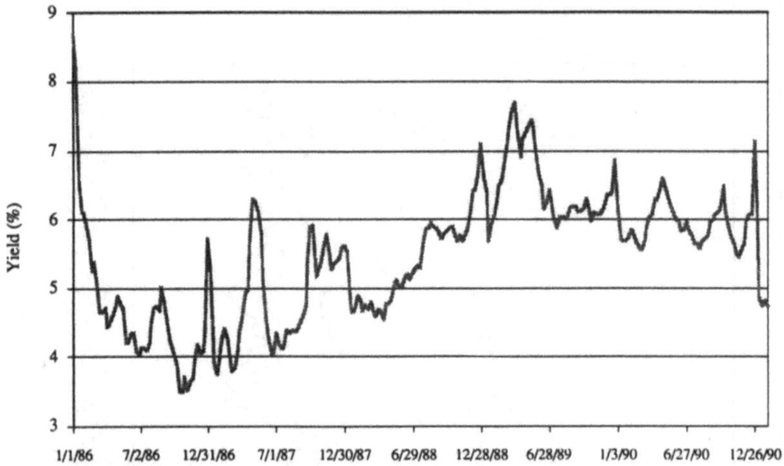

Date	KENNY INDEX
19-Mar-86	4.64
26-Mar-86	4.71
02-Apr-86	4.44
09-Apr-86	4.52
16-Apr-86	4.63
23-Apr-86	4.75
30-Apr-86	4.89
07-May-86	4.78
14-May-86	4.69
21-May-86	4.2
28-May-86	4.19
04-Jun-86	4.32
11-Jun-86	4.35
18-Jun-86	4.07
25-Jun-86	4.02
02-Jul-86	4.14
09-Jul-86	4.13
16-Jul-86	4.08
23-Jul-86	4.21
30-Jul-86	4.5
06-Aug-86	4.71
13-Aug-86	4.73
20-Aug-86	4.68
27-Aug-86	5.01
03-Sep-86	4.79
10-Sep-86	4.52

Date	KENNY INDEX
17-Sep-86	4.24
24-Sep-86	4.17
01-Oct-86	4.02
08-Oct-86	3.87
15-Oct-86	3.49
22-Oct-86	3.48
29-Oct-86	3.7
05-Nov-86	3.51
12-Nov-86	3.63
19-Nov-86	3.71
26-Nov-86	3.91
03-Dec-86	4.17
10-Dec-86	4.05
17-Dec-86	4.1
24-Dec-86	4.36
31-Dec-86	5.72
07-Jan-87	5.25
14-Jan-87	3.91
21-Jan-87	3.74
28-Jan-87	3.9
04-Feb-87	4.25
11-Feb-87	4.42
18-Feb-87	4.27
25-Feb-87	4.06
04-Mar-87	3.78
11-Mar-87	3.83

Appendix 11. *(Continued)*

Date	KENNY INDEX
18-Mar-87	4.1
25-Mar-87	4.35
01-Apr-87	4.58
08-Apr-87	4.93
15-Apr-87	4.98
22-Apr-87	5.62
29-Apr-87	6.31
06-May-87	6.26
13-May-87	6.11
20-May-87	5.79
27-May-87	5.09
03-Jun-87	4.59
10-Jun-87	4.25
17-Jun-87	4.03
24-Jun-87	4.03
01-Jul-87	4.34
08-Jul-87	4.17
15-Jul-87	4.11
22-Jul-87	4.14
29-Jul-87	4.39
05-Aug-87	4.34
12-Aug-87	4.4
19-Aug-87	4.37
26-Aug-87	4.39
02-Sep-87	4.49
09-Sep-87	4.59
16-Sep-87	4.75
23-Sep-87	5.37
30-Sep-87	5.9
07-Oct-87	5.92
14-Oct-87	5.16
21-Oct-87	5.26
28-Oct-87	5.36
04-Nov-87	5.62
11-Nov-87	5.78
18-Nov-87	5.51
25-Nov-87	5.28
02-Dec-87	5.35
09-Dec-87	5.41
16-Dec-87	5.46
23-Dec-87	5.6
30-Dec-87	5.61
06-Jan-88	5.52
13-Jan-88	4.76
20-Jan-88	4.64
27-Jan-88	4.69
03-Feb-88	4.89
10-Feb-88	4.83
17-Feb-88	4.67
24-Feb-88	4.75
02-Mar-88	4.69
10-Mar-88	4.79

Date	KENNY INDEX
17-Mar-88	4.65
23-Mar-88	4.59
31-Mar-88	4.7
06-Apr-88	4.65
13-Apr-88	4.55
20-Apr-88	4.78
27-Apr-88	4.77
04-May-88	4.86
11-May-88	5.07
18-May-88	5.11
25-May-88	5.01
01-Jun-88	5.04
08-Jun-88	5.17
15-Jun-88	5.2
22-Jun-88	5.11
29-Jun-88	5.23
06-Jul-88	5.3
13-Jul-88	5.34
20-Jul-88	5.3
27-Jul-88	5.65
03-Aug-88	5.87
10-Aug-88	5.87
17-Aug-88	5.96
23-Aug-88	5.89
31-Aug-88	5.87
07-Sep-88	5.79
14-Sep-88	5.73
21-Sep-88	5.78
28-Sep-88	5.84
05-Oct-88	5.87
12-Oct-88	5.89
19-Oct-88	5.81
26-Oct-88	5.69
02-Nov-88	5.77
09-Nov-88	5.67
23-Nov-88	5.87
30-Nov-88	6.1
07-Dec-88	6.44
14-Dec-88	6.44
21-Dec-88	6.64
28-Dec-88	7.1
04-Jan-89	6.6
11-Jan-89	6.4
18-Jan-89	5.67
25-Jan-89	5.94
01-Feb-89	6.07
08-Feb-89	6.31
15-Feb-89	6.49
22-Feb-89	6.59
01-Mar-89	6.86
08-Mar-89	7.11
14-Mar-89	7.34

(continued)

Appendix 11. *(Continued)*

Date	KENNY INDEX
22-Mar-89	7.59
29-Mar-89	7.69
04-Apr-89	7.22
11-Apr-89	6.91
19-Apr-89	7.16
26-Apr-89	7.24
03-May-89	7.35
10-May-89	7.45
17-May-89	7.37
25-May-89	6.96
31-May-89	6.65
09-Jun-89	6.5
14-Jun-89	6.15
21-Jun-89	6.25
28-Jun-89	6.43
05-Jul-89	6.15
12-Jul-89	6.02
19-Jul-89	5.87
26-Jul-89	6.05
02-Aug-89	6.05
09-Aug-89	6
16-Aug-89	6.05
23-Aug-89	6.17
30-Aug-89	6.2
06-Sep-89	6.2
13-Sep-89	6.1
20-Sep-89	6.12
27-Sep-89	6.15
04-Oct-89	6.30
11-Oct-89	6.23
18-Oct-89	5.98
25-Oct-89	6.11
01-Nov-89	6.07
08-Nov-89	6.09
15-Nov-89	6.10
22-Nov-89	6.20
29-Nov-89	6.37
06-Dec-89	6.34
13-Dec-89	6.37
20-Dec-89	6.40
27-Dec-89	6.87
03-Jan-90	6.12
10-Jan-90	5.71
17-Jan-90	5.67
24-Jan-90	5.74
31-Jan-90	5.85
07-Feb-90	5.80
14-Feb-90	5.67
21-Feb-90	5.59
28-Feb-90	5.54

Date	KENNY INDEX
07-Mar-90	5.70
14-Mar-90	5.88
21-Mar-90	6.05
28-Mar-90	6.06
04-Apr-90	6.30
11-Apr-90	6.28
18-Apr-90	6.41
25-Apr-90	6.61
02-May-90	6.46
09-May-90	6.38
16-May-90	6.23
23-May-90	6.10
30-May-90	6.00
06-Jun-90	5.97
13-Jun-90	5.82
20-Jun-90	5.85
27-Jun-90	5.98
04-Jul-90	5.81
11-Jul-90	5.78
18-Jul-90	5.65
25-Jul-90	5.64
01-Aug-90	5.58
08-Aug-90	5.67
15-Aug-90	5.72
22-Aug-90	5.75
29-Aug-90	5.98
05-Sep-90	6.01
12-Sep-90	6.08
19-Sep-90	6.10
26-Sep-90	6.20
03-Oct-90	6.49
10-Oct-90	6.12
17-Oct-90	5.92
24-Oct-90	5.75
31-Oct-90	5.63
07-Nov-90	5.53
14-Nov-90	5.45
21-Nov-90	5.54
28-Nov-90	5.66
05-Dec-90	5.97
12-Dec-90	6.07
19-Dec-90	6.28
26-Dec-90	7.14
02-Jan-91	5.97
09-Jan-91	4.86
16-Jan-91	4.74
23-Jan-91	4.82
30-Jan-91	4.73
06-Feb-91	4.49

Appendix 12. Legal Text of Securities and Exchange Commission's Municipal Bond Disclosure Rule

Part 240—General Rules and Regulations, Securities Exchange Act of 1934.

1. The authority citation for Part 240 is amended by adding the following citation:

Authority: Sec. 23, 48 Stat. 901, as amended; 15 U.S.C. 78w.***§240.15c2-12 also issued under 15 U.S.C. 78b, 78c, 78j, 78o, 78o-4, and 78q.

2. By adding §240.15c2-12 as follows:

§240.15c2-12 Municipal securities disclosure.

(a) General. As a means reasonably designed to prevent fraudulent, deceptive, or manipulative acts or practices, it shall be unlawful for any broker, dealer, or municipal securities dealer (hereinafter "Participating Underwriter") to act as an underwriter in a primary offering of municipal securities with an aggregate principal amount of $1,000,000 or more (hereinafter "Offering") unless the Participating Underwriter complies with the requirements of this rule or is exempted from the provisions of this rule.

(b) Requirements.

(1) Prior to the time the Participating Underwriter bids for, purchases, offers, or sells municipal securities in an Offering, the Participating Underwriter shall obtain and review an official statement that an issuer of such securities deems final as of its date, except for the omission of no more than the following information: the offering price(s), interest rate(s), selling compensation, aggregate principal amount, principal amount per maturity, delivery dates, any other terms or provisions required by an issuer of such securities to be specified in a competitive bid, ratings, other terms of the securities depending on such matters, and the identity of the underwriter(s).

(2) Except in competitively bid offerings, from the time the Participating Underwriter has reached an understanding with an issuer of municipal securities that it will become a Participating Underwriter in an Offering until a final official statement is available, the Participating Underwriter shall send no later than the next business day, by first class mail or other equally prompt means, to any potential customer, on request, a single copy of the most recent preliminary official statement, if any.

(3) The Participating Underwriter shall contract with an issuer of municipal securities or its designated agent to receive, within seven business days after any final agreement to purchase, offer, or sell the municipal securities in an Offering and in sufficient time to accompany any confirmation that requests payment from any customer, copies of a final official statement in sufficient quantity to comply with paragraph (b) (4) of this rule and the rules of the Municipal Securities Rulemaking Board.

(4) From the time the final official statement becomes available until the earlier of (i) ninety days from the end of the underwriting period or (ii) the time when the official statement is available to any person from a nationally recognized municipal securities information repository, but in no case less than twenty five days following the end of the underwriting period, the Participating Underwriter in an Offering shall send no later than the next business day, by first class mail or other equally prompt means, to any potential customer, on request, a single copy of the final official statement.

(c) Exemptions. This rule shall not apply to a primary offering of municipal securities in authorized denominations of $100,000 or more, if such securities:

(1) Are sold to no more than thirty five persons each of whom the Participating Underwriter reasonably believes (i) has such knowledge and experience in financial and business matters

that it is capable of evaluating the merits and risks of the prospective investment and (ii) is not purchasing for more than one account or with a view to distributing the securities; or

(2) Have a maturity of nine months of less; or

(3) At the option of the holder thereof may be tendered to an issuer of such securities or its designated agent for redemption or purchase at par value or more at least as frequently as every nine months until maturity, earlier redemption, or purchase by an issuer or its designated agent.

(d) Transactional Exemptions. The Commission, upon written requests, or upon its own motion, may exempt any Participating Underwriter that is a participant in a transaction or class of transactions from any requirement of this rule, either unconditionally or on specified terms and conditions, if the Commission determines that such an exemption is consistent with the public interest and the protection of investors.

(e) Definitions. For the purposes of the rule—

(1) The term "authorized denominations of $100,000 or more" means municipal securities with a principal amount of $100,000 or more and with restrictions that prevent the sale or transfer of such securities in principal amounts of less than $100,000 other than through a primary offering; except that, for municipal securities with an original issue discount of 10 percent or more, the term means municipal securities with a minimum purchase price of $100,000 or more and with restrictions that prevent the sale or transfer of such securities, in principal amounts that are less than the original principal amount at the time of the primary offering, other than through a primary offering.

(2) The term "end of the underwriting period" means the later of such time as (i) the issuer of municipal securities delivers the securities to the Participating Underwriters of (ii) the Participating Underwriter does not retain, directly or as a member or an underwriting syndicate, an unsold balance of the securities for sale to the public.

(3) The term "Final official statement" means a document or set of documents prepared by an issuer of municipal securities or its representatives setting forth, among other matters, information concerning the issuer(s) of such municipal securities and the proposed issue of securities that is complete as of the date of delivery of the document or set of documents to the Participating Underwriter.

(4) The term "issuer of municipal securities" means the governmental issuer specified in section 3(a) (29) of the Act and the issuer of any separate security, including a separate security as defined in rule 3b-5 (a) under the Act.

(5) The term "potential customer" means (i) any person contacted by the Participating Underwriter concerning the purchase of municipal securities that are intended to be offered or have been sold in an Offering, (ii) any person who has expressed an interest to the Participating Underwriter in possibly purchasing such municipal securities, and (iii) any person who has a customer account with the Participating Underwriter.

(6) The term "preliminary official statement" means an official statement prepared by or for an issuer of municipal securities for dissemination to potential customers prior to the availability of the final official statement.

(7) The term "primary offering" means an offering of municipal securities directly or indirectly by or on behalf of an issuer of such securities, including any remarketing of municipal securities (i) that is accompanied by a change in the authorized denomination of such securities from $100,000 or more to less than $100,000 or (ii) that is accompanied by a change in the period during which such securities may be tendered to an issuer of such securities or its designated agent for redemption or purchase from a period of nine months or less to a period of more than nine months.

(8) The term "underwriter" means any person who has purchased from an issuer of municipal securities with a view to, or offers or sells for an issuer of municipal securities in connection with, the offering of any municipal security, or participates or has a direct or indirect participation in any such undertaking, or participates or has a participation in the direct or indirect

underwriting of any such undertaking; except, that such term shall not include a person whose interest is limited to a commission, concession, or allowance from an underwriter, broker, dealer, or municipal securities dealer not in excess of the usual and customary distributors' or sellers' commission, concession, or allowance.

(f) Transitional Provision. If on July 28, 1989 a Participating Underwriter was contractually committed to act as underwriter in an Offering of municipal securities originally issued before July 29, 1989, the requirements of paragraphs (b) (3) and (b) (4) shall not apply to the Participating Underwriter in connection with such an Offering. □

Appendix 13. Municipal Leasing A Fifty State Survey

	ABILITY OF MUNICIPALITIES TO LEASE[1]	SUBJECT TO DEBT LIMITS[2]	PUBLIC BIDDING FOR LEASE[3]	INTEREST RATE LIMITATIONS
Alabama	Yes	1	Yes	None
Alaska	Yes	3	No	None
Arizona	Yes	1	No	None
Arkansas	Yes	1	Yes	None
California	Yes	1	No	None
Colorado	Yes	1	No	None
Connecticut	Yes	3	Yes	None
Delaware	Yes	1	Yes	None
Florida	Yes	1	Yes	Yes
Georgia	Yes	2	Yes	None
Hawaii	Yes	2	Yes	None
Idaho	Yes	2	Yes	None
Illinois	Yes	2	Yes	Yes
Indiana	Yes	1	Yes	None
Iowa	Yes	1	No	None
Kansas	Yes	1	No	Yes
Kentucky	Yes	1	Yes	None
Louisiana	Yes	1	Yes	None
Maine	Yes	1	Yes	None
Maryland	Yes	1	Yes	Yes
Massachusetts	Yes	1	Yes	None
Michigan	Yes	1	Yes	Yes
Minnesota	Yes	3	Yes	Yes
Mississippi	Yes	1	Yes	Yes
Missouri	Yes	1	Yes	Yes
Montana	Yes	2	Yes	None
Nebraska	Yes	1	Yes	Yes
Nevada	Yes	1	Yes	Yes
New Hampshire	Yes	1	No	None
New Jersey	Yes	1	No	None
New Mexico	Yes	1	Yes	None
New York	Yes	1	Yes	None
North Carolina	Yes	3	Yes	None
North Dakota	Yes	2	Yes	Yes
Ohio	Yes	1	Yes	Yes
Oklahoma	Yes	1	Yes	Yes
Oregon	Yes	1	No	None
Pennsylvania	Yes	1	Yes	None
Rhode Island	Yes	1	Yes	None
South Carolina	Yes	1	Yes	Yes

Appendix 13. *(Continued)*

	ABILITY OF MUNICIPALITIES TO LEASE[1]	SUBJECT TO DEBT LIMITS[2]	PUBLIC BIDDING FOR LEASE[3]	INTEREST RATE LIMITATIONS
South Dakota	Yes	1	Yes	None
Tennessee	Yes	3	Yes	Yes
Texas	Yes	1	Yes	Yes
Utah	Yes	1	No	None
Vermont	Yes	1	Yes	None
Virginia	Yes	2	Yes	None
Washington	Yes	1	Yes	None
West Virginia	Yes	1	Yes	None
Wisconsin	Yes	1	No	Yes
Wyoming	Yes	1	No	None

[1]In general, this refers to the police power of a municipality to acquire property for public use, through purchase, gift or lease.
[2]In most cases, the distinction between a lease-purchase obligation, with recourse, and a true lease obligation, without recourse, governs the applicability of constitutional or statutory debt limits;
1—the lease obligation does not constitute a debt, *if* lease rental payments are subject to annual appropriation.
2—the lease obligation would be subject to debt limits, no matter what the lessee's obligation to pay lease rentals may be.
3—lease obligations are not applicable under debt limits.
[3]Public bidding statutes govern the execution of lease obligations by municipalities in those states so indicated.

Source: The Urban Lawyer, Volume 18, Number 1, American Bar Association.

Appendix 14. Range of Fees from AAA Quality to Unrated Quality[1] Size of Financing (Millions of Dollars)

	UNDER 3	3 TO 5	5 TO 10	10 TO 25	25 TO 50	OVER 50
Municipal Bonds:[2]						
Financial Advisor	0.5-1.5%	0.5-1.3%	0.4-1.1%	0.3-0.9%	0.2-0.7%	0.2-0.5%
Underwriter—public bid	1.5-2.3	1.4-2.2	1.3-2.1	1.2-2.0	1.1-1.9	1.0-1.8
negotiated	1.8-3.0	1.6-2.7	1.5-2.5	1.4-2.4	1.3-2.3	1.2-2.2
private sale	1.5-2.8	1.4-2.6	1.3-2.4	1.2-2.3	1.1-2.2	1.0-2.0
Corporate Debt:[2]						
Financial Advisor	1.5-2.5%	1.5-2.5%	1.2-2.0%	1.2-2.0%	1.0-1.5%	1.0-1.5%
Underwriter—public bid	2.5-3.5	2.5-3.5	2.0-3.0	2.0-3.0	1.5-2.5	1.5-2.5
negotiated	3.0-4.5	3.0-4.5	2.5-4.0	2.5-4.0	2.0-3.5	2.0-3.5
limited offering	3.0-4.5	3.0-4.5	2.5-4.0	2.5-4.0	2.0-3.5	2.0-3.5
private offering	2.5-3.5	2.5-3.5	2.0-3.0	2.0-3.0	1.5-2.5	1.5-2.5
Merchant Bank	3.5-4.5	3.5-4.5	2.5-3.5	2.5-3.5	1.5-2.5	1.5-2.5
Corporate Stock:						
Financial Advisor	1.5-2.5%	1.5-2.5%	1.2-2.0%	1.2-2.0%	1.0-1.5%	1.0-1.5%
Underwriter—public offering	3.0-4.5	3.0-4.5	2.5-4.0	2.5-4.0	2.0-3.5	2.0-3.5
limited offering	3.0-4.0	3.0-4.0	2.5-3.5	2.5-3.5	2.0-3.0	2.0-3.0
private offering	2.5-3.5	2.5-3.5	2.0-3.0	2.0-3.0	1.5-2.5	1.5-2.5

Merchant Bank	4.0-5.0	4.0-5.0	3.0-4.0	3.0-4.0	2.0-3.0	2.0-3.0
Venture Capital Firm	5.0-8.0	5.0-8.0	4.0-7.0	3.5-6.0	3.0-5.0	3.0-4.0
Limited Partnerships:						
Financial Advisor	2.5-3.5%	2.5-3.5%	2.0-3.0%	2.0-3.0%	1.5-2.5%	1.5-2.5%
Inv. Banker—public offering	4.0-6.0	4.0-6.0	4.0-5.0	3.5-5.0	3.5-4.0	3.5-4.0
limited offering	3.5-5.5	3.5-5.5	3.5-4.5	3.0-4.5	3.0-3.5	3.0-3.5
private offering	3.0-5.0	3.0-5.0	3.0-4.0	2.5-4.0	2.5-3.0	2.5-3.0
Merchant Bank	4.0-5.0	4.0-5.0	3.0-4.0	3.0-4.0	2.0-3.0	2.0-3.0
Venture Capital Firm	5.0-8.0	5.0-8.0	4.0-7.0	3.5-6.0	3.0-5.0	3.0-4.0
Mortgage Loans:[2]						
Financial Advisor	1.5-2.5%	1.5-2.5%	1.2-2.0%	1.2-2.0%	1.0-1.5%	1.0-1.5%
Inv. Banker—single sale	2.0-3.0	2.0-3.0	1.5-2.5	1.5-2.5	1.0-2.0	1.0-2.0
syndicated	2.5-3.5	2.5-3.5	2.0-3.0	2.0-3.0	1.5-2.5	1.5-2.5
Broker— single sale	2.0-3.0	2.0-3.0	1.5-2.5	1.5-2.5	1.0-2.0	1.0-2.0
syndicated	2.5-3.5	2.5-3.5	2.0-3.0	2.0-3.0	1.5-2.5	1.5-2.5
Bank— single sale	3.0-6.0	3.0-6.0	2.5-5.5	2.5-5.5	2.0-5.0	2.0-5.0
syndicated	5.0-8.0	5.0-8.0	4.5-7.5	4.5-7.5	4.0-7.0	4.0-7.0

[1]Fees are shown as a percentage of the total financing.
[2]Assuming long-term debt, not less than 10 years maturity.

INDEX

MIX
Papier aus verantwortungsvollen Quellen
Paper from responsible sources
FSC® C105338
www.fsc.org

If you have any concerns about our products,
you can contact us on
ProductSafety@springernature.com

In case Publisher is established outside the EU,
the EU authorized representative is:
Springer Nature Customer Service Center GmbH
Europaplatz 3, 69115 Heidelberg, Germany

Printed by Libri Plureos GmbH
in Hamburg, Germany